本项目受国家自然科学基金委员会重大研究计划
"细胞编程与重编程的表观遗传机制"资助

"细胞编程与重编程的表观遗传机制"指导专家组（立项及前期）

组　长：裴　钢
副组长：尚永丰　孙方霖
专　家：朱作言　陈润生　曹晓风　席　真

"细胞编程与重编程的表观遗传机制"管理工作组（立项及前期）

组　长：杜生明
专　家：谷瑞升　江虎军　杜灿平　熊小芸　王岐东

总主编 杨 卫

细胞编程与重编程的表观遗传机制

Epigenetic Mechanisms of
Cell Programming and Reprogramming

细胞编程与重编程的表观遗传机制项目组 编

浙江大学出版社

"中国基础研究报告"
编辑委员会

主 编 杨 卫

副主编 高 文　　高瑞平

委 员

韩 宇	王长锐	郑永和
郑仲文	冯 锋	周延泽
高体玙	朱蔚彤	孟庆国
陈拥军	杜生明	王岐东
黎 明	秦玉文	高自友
董尔丹	韩智勇	杨新泉
任胜利		

总 序

合抱之木生于毫末，九层之台起于垒土。基础研究是实现创新驱动发展的根本途径，其发展水平是衡量一个国家科学技术总体水平和综合国力的重要标志。步入新世纪以来，我国基础研究整体实力持续增强。在投入产出方面，全社会基础研究投入从2001年的52.2亿元增长到2016年的822.9亿元，增长了14.8倍，年均增幅20.2%；同期，SCI收录的中国科技论文从不足4万篇增加到32.4万篇，论文发表数量全球排名从第六位跃升至第二位。在产出质量方面，我国在2016年有9个学科的论文被引用次数跻身世界前两位，其中材料科学领域论文被引用次数排在世界首位；近两年，处于世界前1%的高被引国际论文数量和进入本学科前1‰的国际热点论文数量双双位居世界排名第三位，其中国际热点论文占全球总量的25.1%。在人才培养方面，2016年我国共175人（内地136人）入选汤森路透集团全球"高被引科学家"名单，入选人数位列全球第四，成为亚洲国家中入选人数最多的国家。

与此同时，也必须清醒认识到，我国基础研究还面临着诸多挑战。一是基础研究投入与发达国家相比还有较大差距——在我国的科学研究与试验发展（R&D）经费中，用于基础研究的仅占5%左右，与发达国家15%~20%的投入占比相去甚远。二是源头创新动力不足，具有世界影响力的重大原创成果较少——大多数的科研项目都属于跟踪式、模仿式的研究，缺少真正开创性、引领性的研究工作。三是学科发展不均衡，部分学科同国际水平差距明显——我国各学科领域加权的影响力指数（FWCI 值）在 2016 年刚达到 0.94，仍低于 1.0 的世界平均值。

中国政府对基础研究高度重视，在"十三五"规划中，确立了科技创新在全面创新中的引领地位，提出了加强基础研究的战略部署。习近平总书记在 2016 年全国科技创新大会上提出建设世界科技强国的宏伟蓝图，并在 2017 年 10 月 18 日中国共产党第十九次全国代表大会上强调"要瞄准世界科技前沿，强化基础研究，实现前瞻性基础研究、引领性原创成果重大突破"。国家自然科学基金委员会作为我国支持基础研究的主渠道之一，经过 30 多年的探索，逐步建立了包括研究、人才、工具、融合四个系列的资助格局，着力推进基础前沿研究，促进科研人才成长，加强创新研究团队建设，加深区域合作交流，推动学科交叉融合。2016 年，中国发表的科学论文近七成受到国家自然科学基金资助，全球发表的科学论文中每 9 篇就有 1 篇得到国家自然科学基金资助。进入新时代，面向建设世界科技强国的战略目标，国家自然科学基金委员会将着力加强前瞻部署，提升资助效率，力争到 2050 年，循序实现与主要创新型国家总量并行、贡献并行以至源头并行的战略目标。

"中国基础研究前沿"和"中国基础研究报告"两套丛书正是在这样的背景下应运而生的。这两套丛书以"科学、基础、前沿"为定位，以"共享基础研究创新成果，传播科学基金资助绩效，引领关键领域前沿突破"为宗旨，紧密围绕我国基础研究动态，把握科技前沿脉搏，以科学基金各类资助项目的研究成果为基础，选取优秀创新成果汇总整理后出版。其中"中国基础研究前沿"丛书主要展示基金资助项目产生的重要原创成果，体现科学前沿突破和前瞻引领；"中国基础研究报告"丛书主要展示重大资助项目结题报告的核心内容，体现对科学基金优先资助领域资助成果的系统梳理和战略展望。通过该系列丛书的出版,我们不仅期望能全面系统地展示基金资助项目的立项背景、科学意义、学科布局、前沿突破以及对后续研究工作的战略展望,更期望能够提炼创新思路,促进学科融合,引领相关学科研究领域的持续发展,推动原创发现。

积土成山,风雨兴焉;积水成渊,蛟龙生焉。希望"中国基础研究前沿"和"中国基础研究报告"两套丛书能够成为我国基础研究的"史书"记载,为今后的研究者提供丰富的科研素材和创新源泉,对推动我国基础研究发展和世界科技强国建设起到积极的促进作用。

第七届国家自然科学基金委员会党组书记、主任
中国科学院院士
2017 年 12 月于北京

前 言

表观遗传学是20世纪80年代后期逐渐兴起的一门新学科，是"研究不依赖于DNA序列变化的可继承的性状变化的学科"。多细胞生物如何决定同一基因组在不同细胞类型中的选择性表达，细胞应对生物体内外环境的变化如何进行基因表达的诱导与记忆？这些问题都可以在表观遗传学的研究中找到答案。

表观遗传学研究从最初在动植物中观察到的各种表观遗传现象，到一系列表观遗传修饰的发现，再到大量表观遗传调控因子的鉴定，逐步演化聚焦为染色质对基因表达的调控作用，科学界逐渐认识到表观遗传因子与转录因子共同决定了基因的时空特异性表达。2006年，日本科学家山中伸弥利用转录因子实现了体细胞重编程，我国科学家迅速意识到表观遗传机制在细胞编程与重编程过程中的重要意义，在裴钢、孟安明、陈润生、尚永丰、曹晓风、孙方霖、席真等科学家的倡导下，国家自然科学基金委生命科学部、化学科学部和信息科学部于2008年共同支持了重大研究计划"细胞编程与重编程的表观遗传机制"，该计划在8年里累计资助了包括培育项目、重点支持项目和后期的集成项目共150项，投入经费1.9亿元。该计划积极鼓励前瞻性、原创性和系统性的探索，到2016年底，全面完成了预定的各项科学目标，围绕核心科学问题取得了一系列受到国际瞩目的重大科研成果，如在国际上率先解析 30nm 染色质结构并提出四核小体为染色质重要的结构和调控单元；提出细胞重编程的"跷跷板模型"，并大幅优化了化学小分子诱导体细胞重编程体系；成功建立单倍体胚胎干细胞及半克隆技术，获得"人造精子"；在国际上首先实现从成纤维细胞到肝细胞样细胞的转分化，并逐步走向人工肝应用。在表观基因组研究中取得一系列突破，在国际上首次实现哺乳动物早期胚胎的 DNA 甲基化谱测定、组蛋白甲基化谱测定；解析 DNA 甲基化、去甲基化等多个重要表观遗传修饰酶的活性调控机制等。通过项目的实施培养了一批具有国际水准的优秀科学家，提高了我国表观遗传和细胞命运决定研究的水平，实现了从"跟踪并行"到"全面跻身世界先进水平"的跨越式发展。

本重大研究计划的顺利完成，为后续深入解析表观遗传调控在细胞命运决定中的作用指出了新方向。现有的研究仍大多集中于表观遗传关键因子的发现和功能验证，对重编程细胞中染色质动态变化、细胞表观遗传组学变化及其如何调控转录组变化、决定细胞命运转变的机制等依然知之甚少。我国科学家应继续发挥学科交叉的优势，深入开展表观转录组研究、单细胞转录组及表观基因组研究、表观基因组编辑、四维基因组研究以及跨代表观遗传机制研究，破解以染色质形式存储的遗传信息密码，全面解析表观遗传学在细胞命运决定中的作用，为我国表观遗传研究领域创新能力的全面提升和可持续发展做出不懈的努力。

2018 年 12 月

目 录

第1章
项目概况 ... 01

1.1　项目介绍　　　　　　　　　　　　　　　　01
1.2　研究情况　　　　　　　　　　　　　　　　04
1.3　取得的重大进展　　　　　　　　　　　　　05

第2章
国内外研究情况 ... 07

2.1　以美国为代表的国际表观遗传相关研究计划情况　　09
2.2　表观遗传学的研究现状　　　　　　　　　　10
2.3　表观遗传研究的发展态势　　　　　　　　　14

第3章

重大研究成果　　17

3.1 新表观遗传调控因子和染色质重塑因子及其生物学功能和机制　17

3.2 干细胞自我更新、体细胞重编程的表观遗传学调控机制新发现和动物克隆、生殖技术的重大突破　27

3.3 细胞分化转分化、发育与疾病相关的表观遗传机制研究　37

3.4 构建表观遗传学图谱，揭示胚胎发育表观遗传修饰特点和规律　45

第4章

展　望　　51

4.1 我国表观遗传研究待加强的方向　52

4.2 我国表观遗传研究领域的战略需求　53

4.3 深入研究的设想和建议　53

参考文献　　57

成果附录 63

附录1 重要论文目录 63
附录2 获得的国家科学技术奖项目 125
附录3 代表性发明专利 126
附录4 人才队伍培养与建设情况 129

索 引 141

第1章 项目概况

1.1 项目介绍

"细胞编程与重编程的表现遗传机制研究"（以下简称本重大研究计划）是国家自然科学基金委员会在"十一五"期间启动的一项重大研究计划。本重大研究计划于 2008 年 10 月启动，2016 年底结题，累计资助项目 156 项，其中包括培育项目 68 项、重点支持项目 23 项和集成项目 59 项，申请项目涉及生命、化学和信息等学部，总资助经费达 1.9 亿元。

表观遗传学是 20 世纪 80 年代后期逐渐兴起的一门新学科，研究在 DNA 序列不变的前提下，引起可遗传的基因表达成细胞表型变化的分子机制。表观遗传调控机制是生命现象中一种普遍存在的基因表达调控方式，是调控生长、发育、衰老与疾病发生的重要机制之一。表观遗传调控特别在干细胞维持和自我更新与分化、个体衰老和发育异常，如肿瘤、糖尿病、精神疾病及神经系统疾病等复杂疾病的发生发展中，起着决定性的作用，而且生命个体对环境因素（包括营养、物理化学因素及心理因素等）发生的有序应答在很大程度上依赖于表观遗传调控网络的有效运行。表观遗传调控还在植物发育、植物抗性、植物杂种优势的形成等方面起着重要的作用。

1.1.1 项目部署和综合集成情况

自本重大研究计划组织实施以来，遵循国家自然科学基金委"有限目标、稳定支持、集成升华、跨越发展"的总体思路，围绕总体科学目标下的核心科学问题，共资助培育项目 68 项，重点支持项目 23 项，集成项目 59 项。

表观遗传学在 20 世纪 80 年代后期逐渐兴起，2000 年以后，表观遗传学研究受到广泛重视并成为生命科学研究的前沿和热点。细胞编程与重编程研究囊括了表观遗传学的基本科学问题。2006 年，美国和日本的科学家报道了诱导性多能干细胞的建立，表明体细胞重编程的研究又进入一个新的发展阶段。本重大研究计划启动前，国际上表观遗传学研究已取得重大进展，但科学家对表观遗传机制的了解依然是冰山一角，许多关键问题仍然没有得到解决。这些问题包括 DNA 甲基转移酶如何选择性地作用于靶基因，DNA 去甲基化酶的克隆和鉴定仍然悬而未决，"组蛋白密码"的组成与识别尚需破解，染色质高级结构如何与表观遗传信息互作，非编码 RNA 如何参与表观遗传调控，表观遗传信息可塑性及细胞重编程的分子机制仍不明了，环境、疾病、衰老等与表观遗传调控的关系有待研究，表观遗传调控网络的组成、起源与进化的特点仍不清楚。

本重大研究计划启动之初，该领域是一个新兴领域，我国科学家和国际同行之间研究水平的差距不大，机遇多于挑战，指导专家组加强顶层设计，充分发挥我国从事表观遗传学研究的青年科学人才优势，聚焦国际前沿，开展创新性研究。

围绕科学目标和核心科学问题，本重大研究计划最初部署了以下 5 个研究方向。

（1）表观遗传信息建立和维持的分子机制。

（2）干细胞定向分化过程中的表观遗传机制。

（3）体细胞重编程的表观遗传机制。

（4）组织器官发育与再生过程中的表观遗传机制。

（5）表观遗传信息网络的起源与进化。

本重大研究计划实施4年后，指导专家组在对资助项目的进展情况进行全面调研的基础上，瞄准表观遗传学领域的前沿科学问题和发展趋势，进一步体现本重大研究计划"集中目标和重点突破"策略，把研究方向凝练整合为以下三大集成方向。

（1）DNA甲基化和去甲基化的分子机制及生物学意义。

（2）细胞重编程的表观遗传机制。

（3）细胞重编程过程中核染色质和非编码核酸的高级结构及动态变化。

经指导专家组、项目评审组和承担项目科学家的共同努力，本重大研究计划全面完成了预定的各项科学目标，在三大集成方向均取得了多项具有重大国际影响力的突破性研究成果，带动了我国表观遗传研究总体水平的全面提升和跨越式发展。

1.1.2 学科交叉情况

本重大研究计划实施过程中，指导专家组立足我国表观遗传学研究的现状，着眼国际学科前沿和发展动态，顶层设计，着力推动细胞生物学、生物化学、发育生物学、结构生物学、生物信息科学及临床医学等学科之间的交叉，通过多种形式的项目支持，及时把相关学科的最新思路和技术应用到表观遗传学研究，为实现我国表观遗传学研究的全面跨越式发展做出了突出贡献。

特别值得一提的举措有以下几方面。

（1）加强细胞、生化、遗传、发育、生殖、进化等相关学科的通力合作，在短时间内成功跻身体细胞重编程、细胞核移植及半克隆技术的国际前沿，特别是在单倍体干细胞方面的研究成果在高影响力期刊发表的论文数量超过国际上其他国家发表的论文数量之和，表明我国在该领域处于引领地位。

（2）积极鼓励与物理学交叉，通过引入结构生物学的研究手段，促进了DNA、RNA、组蛋白修饰分子及染色质高级结构的空间结构的解析，有效地推动了关键分子的发现、鉴定及其作用机理的阐明，使我国跃升为该领域的强国。

（3）强调与数理学和信息学的交叉，针对表观遗传学研究中海量数据分析的需求，引入计算生物学的手段，有效推动了表观遗传信号网络互作关键节点的发现和系统生物学机制的阐明。

自本重大研究计划实施以来，遵循国家自然科学基金委"有限目标、稳定支持、集成升华、跨越发展"的总体思路，指导专家组和管理工作组围绕体细胞重编程、个体发育、细胞分化及疾病等相关的表观遗传领域的科学前沿，加强顶层设计，不断凝练科学目标，开展创新性研究，为全面提升我国表观遗传领域基础研究的原始创新能力做出了重大贡献。

1.2 研究情况

1.2.1 总体科学目标

本重大研究计划的科学目标是应用多学科交叉的研究手段，认识细胞编程和重编程过程中表观遗传信息形成、维持和作用的规律和特点，阐明表观遗传调控在细胞生长、发育和环境适应等方面的作用机理，揭示表观遗传网络组成、进化和运行的机制。

1.2.2　核心科学问题

本重大研究计划拟解决以下核心科学问题。

（1）表观遗传信息建立和维持的分子机制。
（2）干细胞定向分化过程中的表观遗传机制。
（3）体细胞重编程的表观遗传机制。
（4）组织器官发育与再生过程中的表观遗传机制。
（5）表观遗传信息网络的起源与进化。

1.3　取得的重大进展

在本重大研究计划执行期间，围绕核心科学问题取得了一系列具有重大国际影响的突破性进展，实现了我国表观遗传研究领域从"跟踪并行"到"全面跻身世界先进水平"的跨越式发展。代表性成果为以下 6 项。

（1）发现新的表观遗传调控因子和染色质重塑因子并揭示其生物学功能和作用机制。

（2）成功建立单倍体胚胎干细胞及半克隆技术，通过操控印记基因成功获得"人造精子"；发现体细胞重编程新的调控机制与方法，提出细胞重编程的"跷跷板模型"。

（3）揭示多种细胞分化转分化的调控机制，特别是发现促进体细胞向肝细胞转分化的关键因子，为生物人工肝的临床应用提供了坚实可靠的基础，发现了与多条疾病相关的表观遗传修饰。

（4）在国际上首次通过采集高通量数据构建表观遗传学图谱，揭示不同物种早期胚胎发育过程中表观遗传修饰的特点和遗传进化的规律，丰富了人们对表观遗传信息网络起源与进化的认识。

（5）在国际上率先解析 30nm 染色质结构，并提出四核小体为染色质重要的结构和调控单元。

（6）首创采用极体基因组移植技术预防遗传性线粒体疾病。

本重大研究计划完成后领域发展态势对比可见表 1.1。

表 1.1 本重大研究计划完成后领域发展态势对比

核心科学问题	计划启动时国内研究状况	计划结束时国际研究状况	计划结束时，我国的研究优势和差距
表观遗传信息建立和维持的分子机制	与国际先进水平差距较大，但是已经有一批年轻科学家回到国内，蓄势待发	国际上在该领域的发展十分迅速，某些领域落后于国内研究水平	DNA 甲基化和去甲基化机制研究、DNA 新甲基化修饰发现与机制研究以及染色质高级结构研究方面处于国际领先水平；其他方面与国际研究水平相当
体细胞重编程的表观遗传机制	国际上体细胞重编程机制研究主要是欧美国家处于绝对领先水平，我国在这方面刚刚起步	国际上体细胞重编程机制研究水平与国内相当，但是在单倍体干细胞研究方面落后于国内	在单倍体干细胞研究方面处于国际最高水平，在体细胞重编程机制研究方面与国际研究水平相当，但是团队数量还是较少
细胞分化转分化、组织器官发育与再生过程中的表观遗传机制	我国在细胞分化与转分化以及器官发育与再生方面的研究水平较国际领先水平差距较大	国际上对其他类型细胞转分化方面特别是如神经细胞、心肌细胞、血液细胞等的研究较多	在肝细胞转分化研究方面处于国际领先水平，其他方面如神经细胞转分化的研究领域与国际水平相当
表观遗传信息网络的起源与进化	我国在该领域的研究基础相对比较薄弱	国际上的研究优势逐渐变小	在早期胚胎发育表观遗传修饰重编程研究中处于国际领先水平，国际上较我国有一定差距

第 2 章　国内外研究情况

表观遗传学的经典定义是"研究不依赖于 DNA 序列变化的可继承的性状变化的学科"。表观遗传现象最早于 20 世纪 30 年代在果蝇中被观察到。自 20 世纪 70 年代起，一系列表观遗传标记被相继发现。随后，大量的表观遗传调控因子得到鉴定，其生物学意义也逐渐明朗。随着多个经典表观遗传现象的生物学机制得到解析，表观遗传学研究的对象也逐渐演化聚焦为染色质结构对基因表达的调控作用。21 世纪初，科学界认识到表观遗传调控机制是生命现象中的一种普遍存在的基因表达调控方式，是调控生长、发育、衰老与疾病发生的重要机制之一。表观遗传调控特别在干细胞维持和自我更新与分化、个体的衰老和发育异常，如肿瘤、糖尿病、精神疾病及神经系统疾病等复杂疾病的发生发展中，起着决定性的作用，而且生命个体对环境因素（包括营养、物理化学因素及心理因素等）发生的有序应答在很大程度上依赖于表观遗传调控网络的有效运行。越来越多的表观调控因子成为新的药物靶标，用于治疗癌症、神经退行性病变等重大疾病。表观遗传调控还在植物发育、植物抗性、植物杂种优势的形成等方面起着重要的作用。表观遗传调控机制主要包括 DNA 甲基化修饰、组蛋白修饰、组蛋白变体、非编码 RNA、核小体定位和染色质高级结构等。

细胞编程与重编程的表观遗传机制

细胞编程与重编程研究囊括了表观遗传学的基本科学问题，是典型的不依赖 DNA 序列变化的可继承的细胞性状变化。2006 年，日本科学家山中伸弥首先报道了利用转录因子实现了体细胞重编程。这一发现使得体细胞重编程的研究进入一个新的发展阶段，集中在三个主要方向：①发现新的因子和技术提高重编程效率和生物安全性；②探索编程与重编程的调控机制；③利用细胞编程与重编程技术，发展针对特定疾病的新型治疗手段。要研究编程与重编程的机理，并向临床和再生医学转化，产生经济效益和社会效益，就需要对表观遗传调控的机制进行深入研究。本重大研究计划启动前，国际范围的表观遗传研究已取得重大进展，但科学家对表观遗传机制的了解依然是冰山一角，许多关键问题仍然没有得到解决。这些问题包括 DNA 甲基转移酶作用于靶基因的选择性，DNA 去甲基化酶的克隆和鉴定，众多组蛋白修饰的组成与识别机制，染色质高级结构如何与表观遗传信息互作，非编码 RNA 如何参与表观遗传调控，表观遗传信息可塑性及细胞重编程的分子机制，环境、疾病、衰老等与表观遗传调控的关系，表观遗传调控网络的组成、起源与进化的特点等。

自 20 世纪初发现表观遗传现象以来，以表观遗传为主题的研究几乎涉及生物体生长发育、个体健康和与环境应答的各个方面。随着时代的变迁，人们的研究手段和认识水平不断进步和发展，表观遗传学的研究侧重点也在逐渐变化。从发现经典的表观遗传现象，到鉴定众多的表观遗传修饰和调控因子，再到综合性表观遗传组学，围绕表观遗传调控机制的研究一直是生命科学的热点领域，同时也是最活跃和最具突破性的热点领域之一。

2.1　以美国为代表的国际表观遗传相关研究计划情况

2003年4月，人类基因组计划宣告完成，美国国家人类基因组研究所（National Human Genome Research Institute, NHGRI）在同年9月启动了DNA元件百科全书（Encyclopedia of DNA Elements, ENCODE）计划，旨在分析和鉴定人类基因组中所有的功能性元件。该项目目前已进行到第四期，其第一期试点研究阶段和第二期技术开发阶段（2003—2007）同时进行，有8个研究组参与试点研究，12个研究组参与技术开发，还有一些研究组参与测序数据的比较、计算和分析，第一、第二期总投入约5500万美元。2007年，ENCODE在 Nature 和 Genome Research 发表了29篇相关论文，报道这一阶段的成果。第三期生产阶段（2007—2017）将研究扩大化和全球化，并建立了数据整合中心和数据分析中心，投入约1.3亿美元。第三期的研究成果于2012年9月以30篇学术论文的形式分别发表在 Nature、Genome Biology 和 Genome Research。ENCODE第四期从2017年2月开始，第一批资助了19个研究项目和2个数据中心，总投入约3300万美元。

美国国家卫生研究院（National Institutes of Health, NIH）于2007年启动了表观遗传组路线图计划（Roadmap Epigenomics Project）。该计划主要有两大目标：①开发全面的综合参考表观基因组图谱，即表观基因组的结构与组织；②发展分析综合表观基因组数据的新技术，即深入理解表观基因组的功能和意义，并发现新的表观基因组的组成成分。路线图计划从以下5个方面展开：①每年提供1000万美元用于建立参考表观基因组图谱中心（Reference Epigenome Mapping Center）；②为一名受资助者提供150万美元/年来发展和运作表观基因组数据分析和协调中心（Epigenomics Data Analysis and Coordination Center），该中心将会支撑参考表观基因组图谱中心，负责传递标准化数据给美国国家生物技术信息中心（National Center for Biotechnology, NCBI）；③资助开发能够明显改善表观遗传研究途径的新技术；④资助研究探索哺乳动物细胞中新的表观遗传标记；⑤资助研究与人类健康和疾病相关的表观遗传变化。在路线图计划的基础上，NIH 还将投入 1.9 亿美元支持 2010 年在巴黎成立的国际人类表观遗传学合作组织（International Human Epigenome Consortium, IHEC）进行的研究。

NIH 于 2015 年宣布启动了 4D 细胞核组计划（4D Nucleome Program），该计划旨在跨学科研究细胞核基因组的结构和组织方式，并注重发展新的尖端技术的改善研究方法。4D 细胞核组计划主要支持以下 6 个方面：①组建细胞核结构与功能跨学科合作组织（Nuclear Organization and Function Interdisciplinary Consortium, NOFIC）；②开发新技术研究染色质高级结构与相互作用；③资助研究亚细胞核精细结构；④发展高通量、高分辨率、高内涵的显微成像技术；⑤成立 4D 细胞核网络组织中心（4D Nucleome Network Organizational Hub），促进合作和资源共享；⑥建立和运作数据协调和整合中心（Data Coordination and Integration Center）。4D 细胞核组计划第一批资助了 29 个为期五年的研究项目，总投入约 2500 万美元。

可以看出，以美国为代表的国际表观遗传相关研究计划主要以大数据组学研究为主，描述性研究成果偏多，同时注重开发相关新技术。

2.2　表观遗传学的研究现状

早期关于表观遗传学的研究主要涉及经典表观遗传现象的发现、染色质修饰与基因表达调控之间关系的确立、表观遗传调控因子的鉴定和纯化，以及利用传统的遗传学和生物化学的方法研究表观遗传调控因子的生理功能等方面。20 世纪 60 年代，人们对多种染色质修饰的功能已经做出了相当有前瞻性的预测。得益于 20 世纪 90 年代以来遗传学与分子生物学技术的发展，研究染色质生物学的技术手段越来越成熟和多样化。同时，表观遗传因子直接调控基因表达的作用最终在真核生物细胞中得到证实。美国洛克菲勒大学的 C. David Allis 与哈佛大学的 Stuart Schreiber 在 1996 年分别克隆和鉴定了组蛋白乙酰化酶和去乙酰化酶，成果分别发表在同一年的 *Cell* 和 *Science* 上。这些进展，使得人们将目光投向种类众多的表观遗传修饰和调控因子。20 世纪 90 年代到 21 世纪初，对各种表观遗传修饰和调控因子的研究出现了井喷式的发展，也正是在对这些表观遗传调控因子作用机理的研究过程中，表观遗传学的理论和方法得以不断完善和进步。近年来，伴随着基因组学和高通量测序技术的发展，本身就处于生命科学前沿的表观遗传学研究进入了全新的发展阶段。

　　本重大研究计划开始于 2008 年，当时国际上表观遗传的研究如火如荼、方兴未艾，特别是新的表观遗传修饰和调控因子的鉴定。由于技术手段日臻成熟，该领域国际竞争激烈，取得的突破性成果较多。以组蛋白甲基化的研究为例，1999 年南加州大学的 Michael Stallcup 克隆了第一个组蛋白精氨酸甲基转移酶；2000 年维也纳生物中心的 Thomas Jenuwein 克隆了第一个组蛋白赖氨酸甲基转移酶；2001 年 Thomas Jenuwein 和剑桥大学的 Tony Kouzarides 两个研究组同时在 *Nature* 独立报道了第一个结合组蛋白甲基化修饰的识别蛋白；2002 年欧美的多个研究组分别解析了组蛋白甲基化识别蛋白的晶体结构和组蛋白甲基转移酶催化结构域的晶体结构。与此同时，国际上在组蛋白去甲基化酶的研究领域展开了激烈的竞争，最终哈佛大学的 Yang Shi 在 2004 年发现了第一个组蛋白去甲基化酶，2006 年北卡罗来纳大学的 Yi Zhang 和哈佛大学的 Yang Shi 分别独立发现了一大类新型的组蛋白去甲基化酶（含 JMJ 结构域）。随后，对组蛋白甲基化的研究逐渐转向对已知因子的生理功能的探索。

　　2006 年日本京都大学的山中伸弥利用转录因子实现了体外的细胞重编程，迅速地在国际上掀起热潮，并推动干细胞领域进入新的研究阶段。日本凭借多年在遗传学和发育生物学研究方面的积累，在表观遗传和干细胞领域有颇多成果。2009年，当时在哈佛大学的Anjana Rao和洛克菲勒大学的Nathaniel Heintz分别独立发现甲基化的DNA在哺乳动物细胞中能够被氧化的现象，发表在同一期 *Science* 上，继而拉开了DNA氧化去甲基化机理研究的序幕。

这些成果的获得，是国际上表观遗传研究人员长期积累、努力研究的结果，也因此在世界上形成了以美国、欧洲和日本为主导的研究格局。以表观遗传为主的研究所和研究中心纷纷成立，研究的主要力量为众多的研究所和大学，例如美国的约翰斯·霍普金斯大学、哈佛大学、美国国家癌症研究所、冷泉港实验室、南加州大学、弗吉尼亚大学、麻省理工学院、洛克菲勒大学、纽约大学、加州大学旧金山分校、加州大学伯克利分校、加州大学洛杉矶分校、加州大学圣地亚哥分校、斯坦福大学、马萨诸塞大学、华盛顿大学（西雅图）、宾夕法尼亚州立大学、圣路易斯华盛顿大学、俄亥俄州立大学等，英国的剑桥大学、牛津大学、爱丁堡大学和约翰·英纳斯中心，德国的马克斯·普朗克研究所，法国的居里研究所，日本的理化学研究所等。在这些机构和大学里，涌现出了一大批表观遗传领域的重要科学家，例如美国国家癌症研究所的Shiv Grewal（异染色质领域）、哈佛大学的Yi Zhang（组蛋白修饰酶领域）、加州大学洛杉矶分校的Steven Jacobson（植物DNA甲基化领域）、马萨诸塞大学的Job Dekker（染色质高级结构领域）、加州大学圣地亚哥分校的Bing Ren（染色质高级结构领域）、剑桥大学的Wolf Reik（哺乳动物DNA甲基化领域）、居里研究所的Edith Heard（X染色体失活领域）和日本九州大学的Hiroyuki Sasaki（基因印记领域）等。

本重大研究计划实施之前，我国在表观遗传领域的研究刚刚起步，正在壮大研究队伍，同时也做出了一些让国际同行认可的研究工作，相关成果发表在 *Cell*、*Nature* 等国际刊物上。代表性的工作包括中国科学院上海生命科学院裴钢研究组对肾上腺激素受体G蛋白偶联受体（G protein-coupled receptor，GPCR）与表观遗传调控的研究，其成果于2005年发表在 *Cell* 上；清华大学孙方霖研究组开展了对组蛋白和表观遗传蛋白对染色质高级结构的调控的研究，成果于2006年发表在 *Genes & Development* 上；中国农业大学巩志忠对DNA去甲基化调控基因沉默的机理研究，成果发表在2006年的 *Plant Cell* 和2007年的 *EMBO Reports* 上。总的来说，当时我国在表观遗传领域的研究与国际上相比起步较晚，团队体量较小。从发表的论文数量来看，Web of Science核心收藏数据库收录的发表于2000年至2008年的表观遗传领域的研究文章和综述共计12542篇，其中近半数为美国学者发表（46.4%），我国学者发表的文章仅占不到5%，国际排名第七。从本重大研究计划实施开始，我国表观遗传学领域的研究得到迅猛发展，2009年至2016年间发表研究文章和综述共计6728篇，占国际上相关论文总数的13.1%（见图2.1），于2014年超过英国。2014—2016年，我国学者每年在学术刊物上发表的表观遗传研究论文数量仅次于美国（见图2.2）。本重大研究计划支持的我国科学家还取得了若干重大研究成果，在世界顶级学术刊物上发表了一系列原创性论文，获得国内外一致赞誉，产生了巨大的国际影响。本重大研究计划实施8年之后，我国在表观遗传学领域无论从论文质量、数量，还是人才队伍建设来看，都已经全面超越日本，成为与美国和欧洲比肩的又一个表观遗传研究中心。

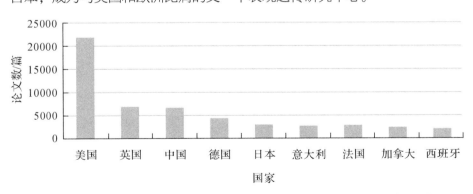

图2.1 2009—2016年Web of Science核心收录的主要国家在表观遗传研究领域的论文数

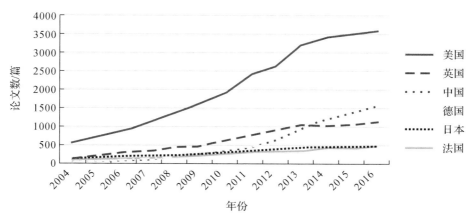

图 2.2　2004—2016 年 Web of Science 核心收录的主要国家在表观遗传研究领域的论文数

2.3　表观遗传研究的发展态势

当前国际上表观遗传研究领域的发展态势可以概括为以下几个方面。

2.3.1　表观遗传调控基因表达机理研究

注重不同表观遗传修饰和调控因子之间的相互作用和调控，同时关注表观遗传修饰因子除酶促反应之外的基因表达调控方式。注重研究细胞利用表观遗传调控机制响应细胞内外环境变化的机理。目前研究对大多数表观遗传修饰酶的催化过程和分子机理基本清楚，深入研究对表观遗传因子的调控机制有助于理解基因表达为什么具有高度的时空特异性，其作用又为什么呈现复杂的动态变化。

2.3.2 新型表观遗传修饰、修饰酶与识别蛋白的发现、鉴定与功能研究

注重生物化学与结构生物学的交叉研究，解析表观遗传修饰酶以及修饰识别蛋白的作用机制。

2.3.3 表观遗传信息建立和维持机制研究

注重与发育生物学和系统生物学的交叉研究，研究不同模式生物中表观遗传信息跨世代继承的载体和机制。注重区分不同要素对于建立和维持表观遗传信息的贡献和调控机理，研究细胞建立和维持时空特异的表观遗传信息的分子机制。

2.3.4 表观遗传调控生殖和发育研究

注重研究表观遗传修饰在这些过程中建立、维持及动态变化的机理。注重表观遗传因子整合信号转导通路、调控细胞分化的分子机制。

2.3.5 表观遗传调控细胞编程与重编程分子机制研究

注重与遗传学和生物信息学的交叉研究，从系统生物学的角度研究细胞命运发生变化的过程中染色质修饰和高级结构的动态变化与调控机理。注重研究这些过程中细胞变化非均一性的机理。注重通过调控表观遗传因子来提高体外细胞重编程的效率。

2.3.6 染色质高级结构和细胞核内亚组织结构研究

注重与物理学和计算生物学的交叉研究，特别注重开发新技术研究染色质高级结构的组成形式和动态变化。注重研究染色质高级结构组成的一般性原则。

2.3.7 表观遗传信息网络的起源与进化研究

注重与生物信息学的交叉研究，构建不同物种胚胎早期发育过程中的表观遗传组学图谱，研究表观遗传信息遗传进化的规律。

2.3.8 与表观遗传调控相关的肿瘤、神经退行性病变等重大疾病研究

注重表观遗传因子参与疾病发生、发展及恶化的调控机理。注重与化学生物学的交叉研究，开发靶向表观遗传因子的潜在的小分子化合药物。

第 3 章 重大研究成果

本重大研究计划的成功实施带动了我国表观遗传学领域研究水平的全面提升和跨越式发展。在发现新的表观遗传调控因子和染色质重塑因子并揭示其生物学功能和作用机制，解析干细胞自我更新、体细胞重编程的调控机理和克隆新方法，揭示表观遗传调控细胞分化转分化、个体发育及疾病发生发展的规律，以及从全基因组的层次上认识表观遗传网络形成和运行的机制等方面，取得了一系列具有重大国际影响的突破性研究成果。

3.1 新表观遗传调控因子和染色质重塑因子及其生物学功能和机制

表观遗传调控机制是生命现象中一种普遍存在的基因表达调控方式。DNA 甲基化修饰、组蛋白修饰、核小体的组装、组蛋白密码的建立与识别、高级染色质结构的建立、维持及转换均是表观遗传调控的重要手段。因此，本重大研究计划针对发现与鉴定 DNA 甲基化修饰、组蛋白修饰的新分子，探讨核小体组装，组蛋白密码的组成、识别和作用的机制，研究染色质高级结构及异染色质与常染色质转换的调控机制及功能，阐明重要信号转导通路在表观遗传信息建立和维持机制中的作用等研究方向予以支持，积极鼓励结构生物学研究，获得了一系列突破性研究成果。

3.1.1 DNA 甲基化与组蛋白甲基化修饰互动调控新机制

DNA 甲基化能引起染色质结构、DNA 构象、DNA 稳定性及 DNA 与蛋白质相互作用方式的改变，从而控制基因表达。参与 DNA 甲基化的酶主要有 DNMT1、DNMT3A 和 DNMT3B。前期的一些研究确定了 DNA 甲基化与组蛋白修饰之间的联系，揭示了一种由组蛋白引导 DNA 甲基化建立的机制。DNMT3A 的 ATRX-DNMT3-DNMT3L（ADD）结构域可以识别未甲基化组蛋白 H3（H3K4me0）。在体外，组蛋白 H3 尾部可以提高 DNMT3A 的酶活性，然而目前尚不清楚其分子机制。

在本重大研究计划支持下，我国科学家确定了 DNMT3A 自抑制形式的 DNMT3A-DNMT3L 和活化形式的 DNMT3A-DNMT3L-H3 复合物的晶体结构，其分辨率分别达到 3.82Å 和 2.90Å。结构和生化分析结果表明，DNMT3A 的 ADD 结构域与催化结构域（catalytic domain，CD）相互作用，通过阻断它的 DNA 结合力，抑制了 CD 的酶活性。组蛋白 H3（而非 H3K4me3）可以破坏 ADD-CD 互作，诱导 ADD 结构域大幅度移动，由此解除 DNMT3A 的自抑制。这些研究结果揭示了 DNA 甲基化的另一个调控层面，确定了 DNMT3A 主要是当存在未甲基化的 H3K4 时在适当的靶位点被激活，并强有力地证实了在整个哺乳动物基因组中 H3K4me3 与 DNA 甲基化之间的负相关性（Guo et al.，2015）。这一研究提供了 DNA 从头甲基化转移酶在最初的基因组定位后意想不到的自抑制和组蛋白 H3 诱导其活化的一些新见解，获得 *F1000* 的推荐。

另外一些研究发现，H3K9 甲基化修饰与 DNA 甲基化修饰的互动机制（Liu et al.，2013），并提出 DNA 维持性甲基化的新模型：通过协同结合甲基化组蛋白及半甲基化 DNA，UHRF1 能够更有效地结合至 DNA 复制叉，进而招募 DNMT1 来完成 DNA 甲基化的维持功能。在小鼠模型中进一步证明了 UHRF1 通过识别 H3K9 甲基化来介导组蛋白修饰与 DNA 甲基化互动的作用，并且明确了 H3K9 甲基化在哺乳动物细胞 DNA 甲基化修饰中起着辅助性而不是决定性的作用。这一工作表明不同物种中 DNA 甲基化修饰与组蛋白修饰互动的程度不同，并且是通过不同的机制来实现的（Zhao et al., 2016a）。我国科学家同时发现 UHRF1/2 可以负调控 DNA 起始性甲基化，提出了 UHRF1/2 一方面促进 DNA 维持性甲基化，一方面抑制起始性甲基化来维持细胞 DNA 甲基化稳态的假说（Jia et al., 2016）。同时，研究发现了半甲基化 DNA（由 DNA 复制后形成）结合表观遗传调控因子 UHRF1 并引起 UHRF1 从闭合到开放的构象改变，从而激活 UHRF1 对组蛋白甲基化修饰的识别，确保 UHRF1 能够精确地定位到基因组的特定区域，发挥维持 DNA 甲基化的功能（Fang et al., 2016）。

3.1.2 DNA 甲基化酶、去甲基化酶活性的调控新机制

近年来，光控蛋白成为研究细胞信号转导时空调控的有力工具。其中，与紫外光激发探针相比，利用双光子激发的探针可以极大地降低细胞的毒性，具有广阔的应用前景。我国科学家以蛋白质化学合成作为核心技术，发展了靶向免疫细胞的光控蛋白探针，并使用新发展的蛋白质探针研究了免疫细胞在精确的时空刺激下的定向运动以及活化的机理。通过发展双光子光控趋化因子探针，我国科学家与合作者研究了活体组织中免疫细胞定向运动的机理，设计并合成了第一例可通过双光子激活的趋化因子探针 hCCL5**。利用该探针的高时空分辨性，在单细胞水平上得出"激酶 PI3K 与 T 细胞对运动方向的感知无关，而与运动的持久性相关"的结论（Chen et al., 2015b）。同时，利用双光子良好的组织穿透性，他们与合作者实现

了在活体组织中（小鼠耳朵及淋巴结）对免疫细胞定向运动的控制。该探针为理解和控制活体组织中细胞定位及定位相关的细胞生物学提供了理想的分子工具。通过发展光控蛋白抗原，揭示了 B 细胞活化早期信号传导动力学原理。通过光控氨基酸筛选技术及蛋白质化学合成技术获得第一例针对 B 细胞的光控蛋白抗原 HEL-K96NPE。我国科学家发现，对于鸡卵溶菌酶（HEL）及其对应的抗体 HyHEL-10（或者表达 HyHEL-10 的 MD4 转基因小鼠的 B 细胞），掩蔽单个位点（HEL 第 96 位的赖氨酸）就足以阻断高达 20pm 的相互作用力和 1800Å^2 的相互作用面。结合高速高分辨率的活体成像技术，利用光控蛋白抗原 HEL-K96NPE，他们实时监测了 B 细胞激活早期突触的形成以及钙流的周期性反应。该探针为深入研究 B 细胞活化早期信号传导动力学过程提供了一个有潜力的工具。

哺乳动物基因组的胞嘧啶上会产生甲基化修饰，称为 5mC（5-甲基胞嘧啶，第 5 种碱基），而 TET 蛋白是哺乳动物细胞中的一种氧化酶，可以执行 DNA 去甲基化功能。哺乳动物 TET 蛋白在受精卵表观遗传重编程、多能干细胞分化、骨髓造血等关键生命过程中扮演着至关重要的角色，其失活也与多种疾病，尤其是血液肿瘤的发生密切相关。以 TET 蛋白为中心的 DNA 去甲基化研究是近年表观遗传学最活跃的领域之一。研究发现，TET 蛋白在去甲基化过程中，将 5mC 氧化为 5hmC（5-羟甲基胞嘧啶，第 6 种碱基）后，可继续催化 5hmC 转化为 5fC（5-醛基胞嘧啶，第 7 种碱基）和 5caC（5-羧基胞嘧啶，第 8 种碱基）。其中，5hmC 在细胞内相对稳定存在，且其含量远远高于 5fC 和 5caC。这一现象一直没有合理的生物学解释。利用结构生物学、生物化学和计算生物学等研究方法，我国科学家揭开了这一谜底。通过结构研究发现，5mC 在 TET 蛋白催化口袋中的取向使得它很容易被催化活性中心俘获并被氧化为 5hmC。5hmC 和 5fC 由于已经有氧的存在，其在催化口袋中被限制，不容易发生进一步的氧化反应，导致 TET 蛋白对这两种碱基活性降低。在这样的催化能力差异下，TET 会很顺利地将 5mC 氧化为 5hmC。而当 5hmC 产生后，TET 又不容易使其进一步氧化为 5fC 和 5caC，这样，细胞内 5hmC 相对稳定，并且其含量远远高于 5fC 和 5caC。这一研究揭示了 TET 蛋白底物的偏好性机制，为基因组中 5hmC 的稳定存在提供了分子水平的解释。在特定的基因区域，TET 蛋白可能被特定的调控因子激活，产生高活性的 TET，将 5hmC 连续氧化为 5fC 和 5caC。这一发现解决了表观遗传学领域的一个难题，也为揭示其他蛋白质逐步催化反应的分子机制提供了新思路和新方法（Hu et al., 2015）。

我国科学家发现锌指结构域蛋白 SALL4A 倾向于结合含有 5hmC 修饰的 DNA。SALL4A 是早期胚胎发育过程中的一个重要基因，它的突变会导致常染色体显性遗传的 Duane-radial ray 综合征。SALL4A 基因敲除的小鼠胚胎在围着床期即停止发育，并很快死亡。研究发现，在小鼠胚胎干细胞中，SALL4A 蛋白主要定位于增强子（enhancer），其与染色质的结合在很大程度上依赖于 TET1 蛋白。进一步分析基因组上 SALL4A 结合位点的胞嘧啶修饰状态发现，这些位点上缺乏稳定的 5hmC，却富集了进一步氧化的产物 5fC 和 5caC，提示 SALL4A 可能促进 5hmC 的进一步氧化。果然，敲除 SALL4 基因后发现在原先的 SALL4A 结合位点上积累了较高水平的 5hmC，敲除 SALL4 降低了 TET2 的稳定结合，不利于 5hmC 的进一步氧化（Xiong et al., 2016a）。这一工作丰富了对 TET 家族蛋白调控的 DNA 氧化和去甲基化过程的理解，并提出了 5mC 的协同性递进氧化概念：5hmC 的结合蛋白 SALL4A 具有细胞特异性和基因组分布区域特异性，它不同于结合 5mC 的 MeCP2 的角色，通过招募 TET2 到特定的位点将 5hmC 进一步氧化成 5fC 和 5caC 从而实现精细（fine-tune）的基因表达调控。这一发现促进了人们对 DNA 甲基化的动态性及其在胚胎干细胞功能及重编程中作用的理解。专业学术评论网站 *F1000* 对该工作进行了介绍和推荐。

以往的一些研究表明 USP7 结合 DNMT1 并通过乙酰化和泛素化作用调控 DNMT1 稳定性，但对于 USP7 介导 DNMT1 稳定性的分子机制目前尚不清楚。为此，我国科学家确定了人类 DNMT1 与 USP7 的晶体结构，分辨率达到 2.9Å。结构和生物化学分析揭示，它们之间的互作主要是通过 DNMT1 的 KG 连接体（linker）和 USP7 从前未被发现的一个酸性口袋介导，这一靠近 USP7 羧基端（C-Terminus）的酸性口袋充当了底物结合位点。这些酸性残基突变可以破坏 DNMT1 与 USP7 之间的互作，促进 DNMT1 转换。KG 连接体赖氨酸残基乙酰化可破坏 DNMT1-USP7 的互作，促进 DNMT1 蛋白酶体降解。采用组蛋白去乙酰化酶（histone deacetylase, HDAC）抑制剂处理可导致乙酰化 DNMT1 上升及总体 DNMT1 蛋白下降。研究人员在一些分化神经细胞和胰腺癌细胞中都观察到了这种负相关关系。这些结果表明，USP7 介导 DNMT1 稳定性受到乙酰化作用的调控，并为设计出靶向 DNMT1-USP7 互作表面的抑制剂提供了结构基础（Cheng et al., 2015）。

3.1.3 组蛋白修饰新密码与解读机制

组蛋白修饰是表观遗传调控基本机制之一，被认为构成一类"组蛋白密码"，调控着遗传信息在染色质层面的解读，在基因表达和细胞命运决定等过程中发挥着重要作用。近年来，众多新型组蛋白修饰被不断发现，其中一大类是组蛋白赖氨酸酰基化修饰，如乙酰化（ac）、丙酰化（pr）、丁酰化（bu）和巴豆酰化（cr）等。关于组蛋白的乙酰化已经有很多研究，而组蛋白巴豆酰化是新近发现的一类从酵母到人类都保守存在的修饰密码，与活跃转录和基因激活密切相关，调控着基因表达及配子成熟等生物学过程。自 2011 年组蛋白巴豆酰化被芝加哥大学赵英明实验室报道（Tan et al., 2011）以来，围绕组蛋白巴豆酰化的产生、消除和识别机制研究成了一个热点。紧跟着的一个学科机遇与挑战则是发现作为"组蛋白密码"直接诠释者的巴豆酰化修饰特异识别阅读器（reader）。

在本重大研究计划支持下，我国科学家发现新型组蛋白巴豆酰化修饰阅读器，先后发表 3 篇高水平的研究论文（Xiong et al., 2016b；Li et al., 2016c；Zhao et al., 2016a），连续报道两类组蛋白巴豆酰化修饰阅读器 YEATS 和 DPF 结构域。通过对表观调控因子 AF9 和 YEATS2 的 YEATS 结构域和 MOZ 的 DPF 结构域的结构与功能研究，发现了 YEATS 结构域和 DPF 结构域都是一类偏好性组蛋白巴豆酰化修饰阅读器，并阐明了该结构域通过特异读取组蛋白巴豆酰化密码促进基因转录的分子细胞机制。值得一提的是，之后有 3 篇重量级的综述性文章专门强调上述关于 YEATS 的重要工作，认为该发现开启了代谢与表观遗传研究的新方向，是在组蛋白修饰调控领域取得的又一重要突破，深化了人们对巴豆酰化修饰生物学的理解和认识。

组蛋白的甲基化是另一种重要的密码。我国科学家发现和证明 KIAA1718（KDM7A）是 H3K9 和 H3K27 的具有双活性的组蛋白去甲基化酶（Huang et al., 2010）；对线虫 ceKDM7A 的研究发现该酶通过结合 H3K4 三甲基而去除 H3K9 和 H3K27 二甲基，揭示了与转录抑制相关的甲基化（H3K9 和 H3K27 二甲基）和与转录激活相关的甲基化（H3K4 三甲基）不在一起的一种机理（Lin et al., 2010；Yang et al., 2010）；完成了含有不同组蛋白甲基化修饰肽段的 6 种 ceKDM7A 蛋白质共结晶，从结构上解释了该酶催化的特异性和底物识别机理；发现 PHF8 是 H3K9 的组蛋白去甲基化酶，该酶位于核仁区，调控 rRNA 转录（Zhu et al., 2010）。相关研究成果以封面文章发表在 *Cell Research* 上，并获专家同期评论及"Sanofy-Cell Research Outstanding Paper Award of 2010"。

另外，鉴于迄今为止鉴定的 20 种组蛋白赖氨酸去甲基化酶只能去除 H3K4、H3K9、H3K27 和 H3K36 这 4 个赖氨酸位点的甲基化修饰，组蛋白上还有多个重要的赖氨酸甲基化位点，包括 H3K79、H4K20 等赖氨酸位点的去甲基化酶还有待发现，精氨酸去甲基化酶仍未找到。我国科学家在本重大研究计划的支持下建立了一套全基因组筛选新型组蛋白去甲基化酶及甲基化调控方式的系统，并用该系统筛选到 H4K20 的新型去甲基化酶 KDM9A/9B，该家族与 LSD 和 JmjC 家族没有保守性，是一类新型的组蛋白赖氨酸去甲基化酶。KDM9A/9B 会在其结合区域调控 H4K20 的甲基化，通过去除 H4K20me1 甲基化激活基因的转录，通过去除 H4K20me3 甲基化激活重复序列的转录，相关研究正在深入进行中。

3.1.4　30nm 染色质纤维高级结构及其动态调控机制

我国科学家利用自主建立的染色质体外组装和冷冻电镜技术，在国际上首次解析了 30nm 染色质纤维的高精度三维冷冻电镜结构（11Å），发现 30nm 染色质纤维是以 4 个核小体为结构单元形成的左手双螺旋结构（见图 3.1），并利用单分子磁镊技术对 30nm 染色质纤维结构建立和调控的动力学过程进行了深入探讨（Song et al., 2014）。*Science* 编辑部以 "Double Helix, Doubled"（双螺旋，无独有偶）为题进行专门介绍。*Science* 同期配有来自 DNA 双螺旋结构模型的发源地——英国剑桥大学的 Andrew Travers 教授 "The 30-nm Fiber Redux"（30nm 纤维的归来）的视点评论。该成果也先后入选最新版的世界著名的生物化学教科书 *Fundamentals of Biochemistry: Life at the Molecular Level* 和 *Lehninger Principles of Biochemistry*。本论文也被 *F1000* 推荐为 "杰出" 级论文。30nm 染色质纤维的三维电镜结构也被德国马克斯·普朗克分子所所长 Patrick Cramer 博士的 "A Tale of Chromatin and Transcription in 100 Structures"（影响染色质和转录研究的 100 项结构研究）（Cramer, 2014）收录。在攻克 30 nm 染色质纤维高级结构这一重大科学问题中，我国科学家取得了重要突破，从而使我国在染色质结构研究领域达到国际领先水平。后续研究建立和完善了描绘全基因组染色质结构的 MNase-Seq 技术——gMNase-Seq（细胞核内染色质结构分析方法）。通过修饰和改造 MNase，提高 MNase-Seq 的空间分辨率，进一步描绘了 30nm 染色质纤维三维结构的动态调控，首次揭示"四核小体串珠"是 30nm 染色质纤维折叠过程中的一个重要中间结构，基因转录过程中组蛋白分子伴侣 FACT 可以负调控"四核小体"结构（Li et al., 2016a）。

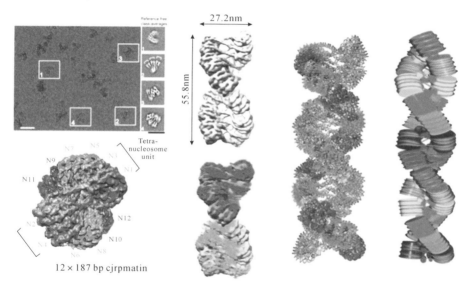

图 3.1　利用冷冻电镜三维重构技术解析 30nm 染色质左手双螺旋高清晰三维结构，解决了分子生物学一个 30 多年悬而未决的重大科学问题

3.1.5　着丝粒染色质建立过程及核小体组装机制

在本重大研究计划支持下，我国科学家系统研究了组蛋白变体 CENP-A 在着丝粒染色质建立过程中的识别、定位及其核小体组装等的作用机制，解析了组蛋白伴侣 HJURP 与着丝粒特征组蛋白变体 CENP-

A-H4复合体的晶体结构，揭示了CENP-A与HJURP的特异识别机制，发现CENP-A关键残基Ser68对HJURP的识别起到重要的调控作用（Hu et al., 2011）。后续的研究工作发现该Ser68的磷酸化修饰调控HJURP与CENP-A的特异性识别，以及CENP-A在细胞周期中的动态组装（Yu et al., 2015）。这些结果为理解着丝粒区核小体的组装机理提供了强有力的证据。被 *F1000* 评述认为"该研究成果揭示出的HJURP特异性识别CENP-A 68位上的丝氨酸将成为日后相关工作的一个焦点"。Patrick Cramer博士也将该工作列入"A Tale of Chromatin and Transcription in 100 Structures"。

3.1.6　组蛋白变体在染色质组装中的作用

组蛋白H3家族包括H3.1、H3.3和着丝粒特异的CenH3，它们在植物中、动物中（从果蝇到人类）都非常保守。拟南芥组蛋白H3.3与H3.1只有4个氨基酸的不同，分别是N端的31和41位以及组蛋白核心区的87和90位。我国科学家通过对拟南芥组蛋白H3.3和H3.1及它们的突变蛋白在核仁rDNA上精细的细胞生物学动态分析，提出和证实了这样一个模型：处于组蛋白H3.3核心区域的87和90位氨基酸介导核小体的组装，而位于N端的31和41位氨基酸则介导核小体的去组装（Shi et al., 2011）。

在研究组蛋白变体H3.3识别和装配的分子机制过程中，我国科学家解析了DAXX-H3.3-H4复合体的晶体结构，揭示了组蛋白变体H3.3被其特异性分子伴侣DAXX和HIRA复合物识别的分子机制，为勾画H3.3的存储途径和理解H3.3的特异识别与组装机制奠定了基础。同时，发现H3.3可以和H2A.Z协同作用，动态调控增强子（enhancer）和启动子（promoter）区域染色质结构，从而维持干细胞自我更新和促进干细胞神经定向分化（Chen et al., 2013）。

3.1.7 表观遗传调控与 DNA 损伤修复互相作用新方式

同源重组（homologous recombination，HR）对于维持基因组稳定和变异至关重要，同源重组之前染色质需要释放出 DNA，而同源重组之后染色质结构需要重新构建。研究发现拟南芥中敲除组蛋白 H2A/2B 分子伴侣核小体组装蛋白 1（NAP1）或 NAP1 相关蛋白（NRP）亚家族成员都会造成植物同源重组频率的下降。组蛋白 H3/H4 分子伴侣 CAF-1 缺失导致的高频率重组现象依赖于 NRP 的存在，且这种依赖在重组中特异，因为 CAF-1 缺失导致的端粒长度缩短的表型不依赖于 NRP。对于 DNA 损伤、组蛋白 H3K56 乙酰化和 DNA 修复基因表达的研究支持了 NAP1/NRP 通过核小体组装/去组装参与同源重组过程。该研究首次证明了组蛋白 H2A/2B 分子伴侣 NAP1 家族成员在真核生物体细胞同源重组中具有重要作用（Gao et al., 2012）。

组蛋白去甲基化酶 KDM5B 属于含 JmjC 结构域的组蛋白去甲基化酶。KDM5B 会依赖于其自身的多聚核糖化和组蛋白变异体的识别在 DNA 损伤位点附近募集，通过其去甲基化酶活性改变损伤区域的染色质结构和状态，进而招募非同源末端连接和同源重组的必要因子 Ku70 和 BRCA1，达到调节 DNA 双链损伤修复的目的，从而维持基因组稳定性。该研究结果有助于解析维持遗传保真性的表观遗传作用，对认识基因组不稳定性相关疾病（如癌症等）具有重要的意义（Li et al., 2014a）。

3.2 干细胞自我更新、体细胞重编程的表观遗传学调控机制新发现和动物克隆、生殖技术的重大突破

体细胞重编程一直是生命科学研究领域的热点之一。体细胞重编程可以通过体细胞核移植技术实现。2006 年日本科学家山中伸弥利用 OSKM 4

个转录因子将体细胞重新诱导成类似于胚胎干细胞的一种细胞类型,即诱导性多能干细胞(induced pluripotent stem cells, iPSC)。这种具有多能干细胞状态的诱导性多能干细胞的建立与发现,不仅开创了体细胞重编程的新方法,更是生命科学历史上的一个里程碑事件。体细胞重编程对再生医学及药物的开发利用,以及治疗人类各种遗传性和功能性疾病都起到了巨大的推动作用。本重大研究计划启动后,指导专家组审时度势,认为体细胞重编程会成为未来前沿研究中的热点,大力支持这一领域的研究,使我国短期内在动物克隆、生殖技术及体细胞重编程机制研究领域取得了丰硕成果,研究水平处于国际领先。

3.2.1 建立单倍体胚胎干细胞及半克隆技术,获得"人造精子"

单倍体干细胞是一种全新的人工细胞类型。在本重大研究计划的支持下,我国科学家在哺乳动物单倍体干细胞的建立和应用上取得了一系列重要进展:首次建立能够替代精子完成生殖过程的小鼠和大鼠孤雄单倍体胚胎干细胞系,以及猴孤雌单倍体胚胎干细胞系;开发了基于单倍体胚胎干细胞的遗传筛选和遗传修饰技术;将单倍体干细胞作为新的技术和工具应用于生殖发育生物学研究,获得了来自两只雌性"同性"小鼠的后代;首次创建了一类新型细胞类型——哺乳动物异种杂合二倍体干细胞,丰富了生殖发育研究理论和体系,拓展了生殖发育研究的范畴,展示了单倍体干细胞可能为生殖生物学、发育生物学、遗传学和进化生物学带来的重要应用前景。相关研究发表在 *Nature*、*Cell*、*Cell Stem Cell* 等刊物上,其中在高影响力期刊发表的论文数量超过国际上其他国家发表的论文数量之和,这表明我国在单倍体干细胞技术及其相关研究领域处于引领地位。

（1）建立哺乳动物单倍体干细胞系并开展其特性与功能研究

我国科学家利用核移植技术，将精子注射到去核卵母细胞中，构建了孤雄发育的单倍体胚胎，并从中提取建立了孤雄单倍体胚胎干细胞系（ahES cells）。这些细胞具有分化形成三胚层细胞以及生殖细胞的能力，并且在注射入 M Ⅱ 卵母细胞质内后可以替代精子使卵子受精，胚胎可进一步发育成健康可育的个体——半克隆小鼠，即半克隆技术。他们将 PGK 启动子控制的 neor 抗性基因电转进 ahES 细胞系 AHGFP-4 中，并进行 G418 药物筛选，建立了转基因大单倍体干细胞系。他们挑取了 3 个转基因系进行 ICAI 实验，发现 3 个系都能产生到期胎儿，其中 2 个系分别获得 7 只和 1 只健康存活转基因动物。这一研究提供了一种新的研究生殖和基因印记等基础问题的模型，提供了一种新的获得转基因动物的方式，也为辅助生殖新技术的开发提供了新的思路。相关成果于 2012 年发表于 *Nature*（Li et al., 2012a），并入选当年"中国十大科学进展"。建立单倍体上胚层干细胞系（hEpiSCs），为单倍体干细胞遗传筛选和二倍体维持等发育机制研究提供新的工具（Shuai et al., 2015）。

（2）利用单倍体干细胞开展生殖发育研究

利用单倍体胚胎干细胞创造出哺乳动物异种杂合二倍体胚胎干细胞。我国科学家通过细胞融合技术将小鼠孤雄（雌）和大鼠孤雌（雄）单倍体胚胎干细胞融合，从而绕开了小鼠和大鼠的精卵融合后无法发育的生殖隔离障碍，获得了异种杂合二倍体胚胎干细胞（Li et al., 2016b）。这类杂交细胞具有胚胎干细胞的三胚层分化能力，甚至能分化形成早期的生殖细胞，并在培养和分化过程中保持异种二倍体基因组的稳定性。基因表达分析发现，异种杂合二倍体细胞展现出"高亲""低亲"等独特的基因表达模式及独特的生物学性状，对两者结合进行分析能有效挖掘出物种间性状差异的分子调控机制。同时，杂合细胞的 X 染色体失活也不采用哺乳动物常见的"随机失活"模式，而是采用小鼠 X 染色体特异失活模式。利用该特性，研究系统鉴定了小鼠 X 染色体失活逃逸基因。这是首例人工创建的、以稳定二倍体形式存在的异种杂合胚胎干细胞，为进化生物学、发育生物学和遗传学等研究提供了新的模型和工具，相关成果发表于 2016 年的 *Cell*。

我国科学家利用单倍体融合细胞培育出健康可育小鼠，证实囊胚期之前的父母源基因组的互作对胚胎发育不是必需的，为发育生物学和辅助生殖技术开发提供了新的思路（Li et al., 2015）。对单倍体胚胎干细胞进行精确的印记修饰实现了"同性"生殖（Li et al., 2016c）——印记基因是在父源和母源染色体上具有不同表达模式的基因。单倍体干细胞有来自于父源或母源单方面的遗传物质，同时具备胚胎干细胞在基因操作上的优势，是合适的印记基因研究平台。利用孤雌胚胎干细胞为平台，通过敲除两个父源甲基化的印记调控区，发现可以修复该区域的印记基因表达模式，再将敲除的孤雌单倍体干细胞注射到卵母细胞中，获得了具有两个母亲的孤雌胚胎并能够以较高的效率获得孤雌来源的小鼠。这一研究给哺乳动物的孤雌生殖提供了新的途径，为动物繁殖和育种相关研究提供了新的思路，同时也为印记异常和治疗方式的探索提供了新的研究方法。

（3）利用单倍体干细胞技术进行遗传筛选和遗传修饰

单倍体细胞的"受精"能力随着细胞的传代逐渐丢失，特别是经过基因编辑后，这些细胞再注入卵子中很难获得健康的半克隆小鼠。在本重大研究计划的支持下，通过将调控雄性印记基因 H19 和 Gtl2 表达的 H19-DMR 和 IG-DMR 敲除后获得了能稳定产生半克隆小鼠的"类精子细胞样"的单倍体细胞（"人造精子"），并证明它们能稳定高效地支持获得遗传修饰的半克隆小鼠。重要的是，这些"人造精子"细胞携带 CRISPR-Cas9

文库，能够进一步产生大量的携带不同突变基因的小鼠，从而为开展小鼠个体水平的遗传筛选和遗传修饰提供了技术支持（Zhong et al., 2015）。进一步证明卵子来源的单倍体细胞在去除 H19-DMR 和 IG-DMR 后，同样具备"人造精子"的能力，从而在哺乳动物中实现了高效的孤雌发育。另外，还建立了来自人卵母细胞的孤雌单倍体胚胎干细胞（Zhong et al., 2016）。

在单倍体干细胞中，CRISPR-Cas9 能被高效用于遗传疾病治疗。通常将 CRISPR-Cas9 直接注入携带白内障遗传缺陷的受精卵中进行基因编辑，可以治愈小鼠的白内障遗传缺陷，而这种通过直接胚胎注射的方法存在两个问题：一是新生小鼠被治愈的概率较低，约为 30%；二是存有少量实验脱靶现象（Wu et al., 2013）。为解决这些问题，我国科学家从白内障小鼠的睾丸中获取了携带纯合遗传突变的精原干细胞，并通过 CRISPR-Cas9 修复遗传缺陷，建立了一系列来自单个精原干细胞的细胞系。随后，对这些细胞系进行深入分析，将满足突变位点均已修复，不存在脱靶问题并维持正常的表观遗传特性的"优质"细胞移植到体内，获得了完全健康的小鼠。这一研究成果为人类基因治疗提供了一条新的思路（Wu et al., 2015）。论文发表后，*Cell* 做了全球新闻同步发布，得到 *Cell Stem Cell*、*Nature*、*Nature Review Genetics*、*Addgene Blog*、*Sci Bull*、*MIT Technology Review*、*F1000* 等杂志和网站的点评或专评 7 次，认为这一研究显示了"CRISPR-Cas9 技术能够用于人类遗传疾病的治疗"，掀起基因治疗的热潮。两篇论文均进入 ESI（Essential Science Indicators）全球 1% 论文（Percentiles for Papers Published by Field, 2006—2016）。

3.2.2 首创极体基因组移植预防遗传线粒体疾病技术

我国科学家将极体作为受体基因放入健康供体卵母细胞的胞质中，实现了基因组转移——线粒体置换（Wang et al., 2014b）。这一研究从干细胞治疗的细胞替代技术推进到细胞器置换技术，为难治性疾病的治疗提供了新的策略和路径。成果论文发表在 Cell 上。Nature 和 Nature Review Genetics 发文做了题为"阻断遗传病变异传递"和"聚光灯下的线粒体置换技术"的介绍，并称该发明"重要的是证明极体移植的可行性，并显著提高了线粒体移植疗法的效率"。美国医学遗传学会主席 Herman 赞扬该研究成果：为线粒体疾病干预提供了有趣的假设、崭新的发现、先进的干预手段。英国人类受精和胚胎管理局（Human Fertilisation and Embryology Authority, HFEA）特别发 45 页专文综述《极体移植防止线粒体病的安全性和有效性》，供英国公众和议会讨论修改法律允许"线粒体 DNA 置换"参考，并认为中国科学家研发的极体移植防止线粒体病技术比英美科学家使用的原核移植（pronuclear transfer, PNT）、母系纺锤体移植（maternal spindle transfer, MST）有五大先进性：①降低线粒体 DNA 带进子代细胞的携带量；②与纺锤体移植相比，减少了将染色体遗留在细胞内的风险（而极体中染色体都存在）；③避免应用细胞支架抑制因素从受精卵或卵母细胞中取出原核或纺锤体；④避免应用更传统的操作方法，可以减少损伤患者细胞核和减少损伤供体卵细胞的机会，因此能极大地提高效率；⑤可以同时实施第一极体移植、第二极体移植、原核移植、纺锤体移植，因此能成倍提高成功率。

过去，我国在线粒体置换技术研发这一国际生物高科技竞争领域的研究相对空白，极体移植预防线粒体置换技术的发明使得我国在该领域的研究不仅进入国际先进行列，而且在技术创新上有独特的贡献。该研究成果获得国家科学技术进步奖二等奖，并入选 2014 年度"中国科学十大进展"。

3.2.3 提出谱系决定因子诱导成体细胞转变成 iPSC 的"seesaw 模型"

传统的 iPSC 技术是通过向成体细胞中导入多潜能性相关基因而建立

多潜能性。我国科学家研究发现，谱系决定因子可以替代多潜能性基因，诱导成体细胞转变成 iPSC。基于这一新发现，提出了"seesaw 模型"，从新的角度解释了细胞命运决定的理论。"多潜能性"是各种不同的分化谱系之间达到平衡所形成的细胞状态，通过不同细胞谱系之间的平衡诱导重编程的发生可能是一个普适的原理：①首次发现作为谱系分化因子的整个 GATA 家族蛋白的 6 个成员都能替换 Yamanaka 四因子中的 OCT4。此外，证实了 GATA 家族的所有成员都能够抑制外胚层分化基因的表达，验证了之前的"跷跷板"模型。随后进一步发现 GATA 转录因子通过中间桥梁因子 SALL4 来建立起多能性状态所需的干性网络（Shu et al., 2015）。②在之前建立的完全使用小分子化合物实现体细胞重编程为多潜能干细胞（CiPS）的方法中，发现在小分子诱导重编程过程中也存在着动态"seesaw 模型"——与传统转录因子诱导重编程截然不同，在重编程早期，小分子启动 XEN 相关基因的表达，使"seesaw"倾斜并使细胞进入 XEN-like 状态；在重编程后期，通过切换至小鼠 ES 细胞培养条件诱导 Sox2 的表达，将倾斜的"seesaw"重新恢复到平衡位置，从而使细胞进入到多潜能状态（Zhao et al., 2015b）。这些结果表明"seesaw 模型"是个普适性的改变细胞命运的模型。

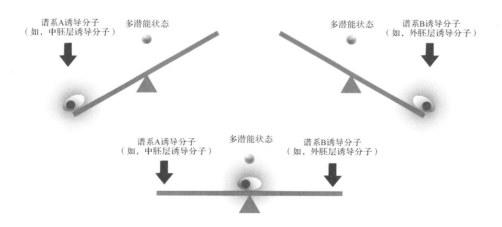

图 3.2 "跷跷板模型"为普适性的改变细胞命运的模型

3.2.4 传统 Yamanaka 因子诱导 iPSC 的方法的安全性评价

之前的体细胞核移植研究表明，人们可以通过连续的核移植技术将体细胞进行连续的重编程，并且每一代都能产生可存活的个体。但是作为体细胞重编程的另一种经典方法——即 iPSC 诱导，却尚未报道过是否能够进行连续诱导重编程以及能进行几次。同时，自 iPSC 诞生起，科学家们就对其应用到临床上的安全性和可行性进行着全方位的评估，是否存在基因突变以及是否对发育潜能构成影响。这不仅关系到细胞的质量，而且影响未来应用的安全性。我国科学家在国际上首次利用 Tet-on 系统建立了 OSKM 介导的联系六代体细胞重编程系统，发现随着连续重编程的进行，诱导性多能干细胞小鼠的活性变得越来越差，其中主要原因是随着突变代数的增加而逐步积累。该发现进一步提示传统方法诱导产生的 iPSC 可能具有一定的风险。成果发表在 2015 年的 *Nature Communications* 上。这一成果也被 *Cell Stem Cell* 等杂志上的多篇论文引用。

3.2.5 深入探讨体细胞重编程调控机制，发现新调控分子

尽管 Yamanaka 因子诱导 iPSC 的方法被广为采用，但人们对于 Yamanaka 因子在重编程细胞中的动态结合模式以及其结合如何调控转录组变化，决定细胞命运转变的机制等依然知之甚少。在本重大研究计划支持下，我国科学家结合了二次重编程系统和高通量测序技术，以核心重编程因子 Oct4 为例开展了系统性研究。他们通过分析体细胞重编程过程中 Oct4 在基因组上的结合，揭示了：在体细胞重编程的不同阶段，Oct4 分层次地结合到基因组上，其结合受到了基因组上预先设置的表观遗传修饰的影响；Oct4 的结合对于重编程过程中多能网络的层次性有序激活具有重要作用（Chen et al., 2016a）。另外，还发现 Oct4 等多能因子可通过结合干细胞特异

的启动子使普遍表达的基因产生干细胞特异的转录本（Feng et al., 2016）。

我国科学家发现了 N6-甲基腺嘌呤（m6A）RNA 甲基化受到 microRNAs 的调节并促进体细胞重编程。真核生物的 RNA 分子上可发生 100 多种修饰，其中 m6A 的甲基化修饰是高等生物 mRNA 上含量最为丰富的修饰。m6A 修饰参与调控 mRNA 的剪接、运输、稳定性和翻译效率等，并且与肥胖和肿瘤等多种生理功能异常及疾病相关。我国科学家绘制了小鼠胚胎干细胞（embryonic stem cell, ESC）、诱导性多能干细胞（iPSC）、神经干细胞（neural stem cell, NSC）和睾丸支持细胞（sustentacular cell, SC）转录组的 m6A 修饰图谱，发现了 m6A 修饰在多能性细胞与分化的细胞间的分布差异。生物信息学分析提示 m6A 修饰区域富集的特征序列具有与 microRNA 的序列互补配对的偏好性。多层次的细胞与分子生物学实验证明，microRNA 可以通过序列互补的方式，引起 mRNA 相应位点区域 m6A 修饰的产生。提高 m6A 修饰水平可以提高 Oct4 等关键多能性调控基因的表达量，促进小鼠成纤维细胞重编程为诱导性多能干细胞（Chen et al., 2015a）。该研究成果揭示了 microRNA 通过序列互补调控 mRNA 甲基化修饰形成这一全新的作用机制，发现了 m6A 修饰在促进体细胞重编程为多能性干细胞中的重要作用，在解析 m6A 修饰形成的位点选择机制，拓展 microRNA 的新功能和发现新的细胞重编程调控因素方面均取得了开创性的重大突破。研究成果以封面文章发表在 *Cell Stem Cell* 上，入选当期特写报道和 *Cell* 出版社的研究特色论文（Stem Cell Highlights from Cell Press）及 *Abcam* 月度优秀表观遗传学研究论文。

我国科学家发现 E3 泛素化酶 RNF20 泛素化 H2B 调控了细胞重编程早期阶段染色体结构的紧凑程度进而影响基因转录因子的招募，从而影响了重编程早期阶段的多能性基因的起始表达。当 RNF20 缺失后，细胞重编程的早期阶段的多能性基因网络建立失败。另外是在 iPSC 中，RNF20 通过泛素化 H2B 调控了复制过程中的 DNA 损伤反应过程。当 RNF20 缺失后，诱导性多能干细胞在复制过程中产生的 DNA 断裂不能及时进行同源重组修复，从而造成其基因组不稳定，最终导致细胞凋亡。生殖细胞特异敲除 Rnf20 导致减数分裂的异常与雄性不育（Xu et al., 2016）。

3.2.6　发现维持干细胞自我更新及发育多能性的新信号

ESC 在体外可无限自我更新和分化为机体内任何种类的细胞，在器官再生和细胞替代治疗中具有广阔的应用前景。然而，人 ESC 维持自我更新及发育多能性的分子调控机制尚有很多问题不清楚，妨碍了将其分化的细胞安全有效地应用于临床。因此，对人 ESC 如何维持其自身特性的机制进行深入的研究显得尤为重要。

我国科学家利用人全基因组范围转录因子 siRNA 文库筛选了参与人 ESC 自我更新的转录因子，发现一系列对维持人 ESC 特性具有重要作用的基因。研究发现，PHB 在维持人 ESC 正确的组蛋白甲基化修饰方面发挥着独特的作用，从而维持人 ESC 自我更新和促进人成体细胞的重编程过程。进一步的研究发现，PHB 可以和组蛋白变异体 H3.3 的伴侣蛋白 HIRA 复合体相互作用，并维持 HIRA 复合体成分的蛋白质稳定性。此外，在人 ESC 中，PHB 和 HIRA 共同调控全基因组范围内 H3.3 在染色质上的富集，特别是参与调控的 H3.3 在异柠檬酸脱氢酶（isocitrate dehydrogenases，IDHs）基因启动子区域的富集及 IDH 基因的表达，从而控制对 ESC 命运具有重要作用的关键代谢产物 a-酮戊二酸（a-ketoglutarate，a-KG）的产生，进而塑造正确的组蛋白甲基化水平，维持人 hESC 自我更新和表观遗传学特性。基于这一结果，我国科学家提出了人 ESC 特性维持的表观–代谢调控环路（Zhu et al., 2017）。

组蛋白去甲基化酶在发育多能性的建立和维持中发挥着关键作用，但在这些过程中具有重要作用的组蛋白去甲基化酶并没有被全部鉴定出来，而且组蛋白去甲基化酶在此过程中发挥作用机制仍不是十分清楚。我国科学家揭示了 H3K9 去甲基化酶 Jmjd1c/Kdm3c 通过调控 miR-200 和 miR-290/295 家族的表达来抑制 MAPK/Erk 信号通路激活和 EMT 发生，进而促进胚胎干细胞自我更新以及发育多能性的维持和建立。在体细胞重编程过程中，干扰或者敲除 Jmjd1c 的表达将显著降低重编程效率，而过表达 miR-200 家族则可以部分地恢复 Jmjd1c-null MEF 的重编程效率。这为发育多能性的建立与维持的机理提供了新视角，也为多能细胞在细胞治疗与再生医学上的应用奠定了理论基础。

BMP 信号通过下游的 Smad 蛋白在转录水平直接上调去磷酸化酶 DUSP9 的表达，从而促使 ERK 蛋白去磷酸化并失去活性，而维持较低的 ERK 活性有利于细胞处于自我更新而不分化的状态，因此，在 BMP 和 LIF 的共同作用下，ERK 活性被维持在合适的水平从而使小鼠胚胎干细胞维持在自我更新而不分化的状态。该研究成果对于理解小鼠胚胎干细胞命运决定的分子机制有着重要的影响，为再生医学的进一步研究奠定了基础（Li et al., 2012b）。研究成果发表后，*Nature Chemical Biology* 对此撰文介绍，*F1000* 进行了评述。

3.3 细胞分化转分化、发育与疾病相关的表观遗传机制研究

3.3.1 发现促进体细胞向肝细胞转分化的关键因子

我国科学家发现慢病毒系统在成体小鼠鼠尾成纤维细胞中过表达肝脏细胞转录因子 FOXA3、HNFLA、GATA4 这 3 个转录因子可以成功

将 p19 失活的小鼠鼠尾成纤维细胞转分化为肝细胞样细胞（hepatocyte-like cells, iHep），同时发现 p53 的激活是去分化重编程的关键抑制机制（见图 3.3）。iHep 细胞不仅能整合到小鼠肝脏中，还在其体内发挥了肝实质细胞的功能。此外，移植 iHep 细胞的 Fah-/- 小鼠和 NOD/SCID（非肥胖糖尿病/重症联合免疫缺陷）小鼠均没有肿瘤形成，进一步说明了 iHep 细胞的安全性（Huang et al., 2011）。利用 FOXA3、HNF1A、HNF4A 三个转录因子和 SV40 Large T 可将人胚胎成纤维细胞诱导转分化为可增殖的功能肝细胞（human hepatocyte-like cell，hiHep）。hiHep 具有与人类原代肝细胞相似的基因表达谱，也获得了肝脏的体外功能，尤为重要的是，hiHep 具有良好的胆汁排泄能力，可应用于药物开发过程中药物胆汁排泄能力的评估。相关工作获得 *Cell Stem Cell* 专评和 *SciBX* 专评，它们认为 hiHep 的获得克服了分化细胞不能增殖的缺陷，认为接转分化获得的功能肝细胞向药物研发和治疗所需肝细胞迈出了令人振奋的一步；*F1000* 推荐并评论本工作：此研究可以大规模获得功能肝细胞，对于整个领域来说是十分重要的进展。该工作被 *Cell* 评为"2014 中国年度论文"。

进一步研究成功将产生的功能肝细胞（hiHep）扩增至临床数量级，与生物人工肝装置结合，成功救治了急性肝衰竭的小型猪（Shi et al., 2016）。被救治的小型猪存活率显著高于对照组，并且各项血生化指标和炎症指标都恢复到正常水平，为 hiHep 的临床应用提供了坚实可靠的基础。2016 年 1 月，我国科学家与南京鼓楼医院合作，成功进行了第一例急性肝衰竭患者的救治实验，成功救治一位罹患乙肝 40 年并突发急性肝衰竭的患者。患者在接受 hiHep 生物人工肝治疗后各项肝功能指标均明显好转，且无任何不良反应。这一病例标志着新型 hiHep 生物人工肝第一例临床治疗成功完成。推进 hiHep 生物人工肝的临床应用将极大地促进我国科研成果转化的进程。

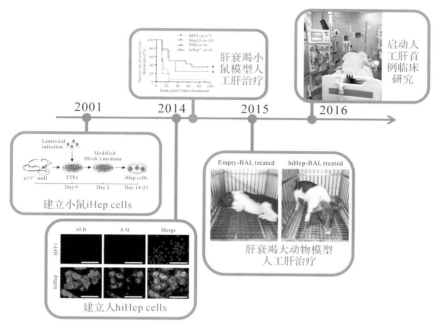

图 3.3 体细胞向肝细胞转分化及人工肝治疗肝衰竭

3.3.2 发现诱导胚胎干细胞或体细胞向神经分化、转分化的表观遗传学机制和细胞信号通路

组蛋白的修饰形式与基因的转录调控密切相关，在神经发生中扮演着重要角色。在胚胎干细胞神经分化过程的上胚层阶段，抑制组蛋白去乙酰化酶活性可以显著抑制胚胎干细胞的神经分化，并促进中内胚层分化，提示组蛋白去乙酰化在神经发育中的重要作用。该研究发现，HDAC1 特异地结合到 *Nodal* 基因区域，组蛋白去乙酰化酶抑制剂能显著上调 *Nodal* 的表达。*Nodal* 信号的活化促进中内胚层命运的表型、抑制神经命运决定，提示 HDAC1 通过拮抗 *Nodal* 维持上胚层阶段的神经命运决定（Liu et al., 2015）。在胚胎干细胞体外神经分化的 0~4 天，H3K9 乙酰化水平逐步下

调；在分化的 4~8 天，其水平逐步上调。相应地，伴随着神经分化的进程，H3K9 乙酰化在全能性相关基因的基因组上的富集减少，在神经发育相关基因的基因组上的富集显著增加。组蛋白去乙酰化酶抑制剂能通过抑制 HDAC3 的活性促进全能性基因的表达并维持人胚胎干细胞的全能性。在神经分化过程中，组蛋白去乙酰化酶抑制剂能拮抗 HDAC1/5/8 的抑制效应，从而促进神经发育相关基因的表达（Qiao et al., 2015a）。组蛋白的甲基化修饰亦参与神经分化调控。在研究中发现并证明 KIAA1718（KDM7A）是 H3K9 和 H3K27 的具有双活性的组蛋白去甲基化酶，通过对 H3K9me2 和 H3K27me2 的去甲基化修饰正向调控 *Fgf4* 基因的转录水平，从而调控胚胎干细胞向神经细胞分化，并在鸡胚体内证明该酶与神经发育相关（Huang et al., 2010）。同时发现，DNA 的去甲基化在神经发生中起重要作用。AF9 基因对于人胚胎干细胞神经分化和神经发育相关基因的转录激活是充分且必要的，我国科学家发现 DNA 羟氧化酶 TET2 能够与 AF9 相互作用，并且 AF9 和 TET2 共同定位于 5hmC 阳性的神经细胞中。AF9 通过识别含有 AAC 的作用原件并结合靶基因启动子区域，招募 TET2 结合到 AF9 与 TET2 的共同下游基因，介导 5mC 向 5hmC 的转变，并激活这些神经发育关键因子的表达，从而促进人胚胎干细胞的神经命运决定（Qiao et al., 2015b）。

  利用神经干细胞进行神经系统功能障碍的治疗是当今再生医学领域的研究热点，诱导成体细胞向神经干细胞的转化，预示着体细胞同样可以不经过多能性状态，直接转变为神经干细胞样的成体干细胞，为细胞治疗提供了丰富的细胞来源。我国科学家成功利用特定因子（*Pax6*、*Ngn2*、*Hes1*、*Id1*、*Ascl1*、*Brn2*、*c-Myc* 和 *Klf4*）诱导睾丸支持细胞转分化为神经干细胞，这些诱导获得的神经干细胞（induced neural stem cells，iNSCs）表达与正常神经干细胞一样的分子标记物，整个基因组的表达情况也与正常的神经干细胞高度相似。功能上，iNSCs 可以维持自我更新并分化成各种有电生理功能的神经元，如多巴胺能神经元、γ-氨基丁酸神经元和乙酰胆碱能神经元。重要的是，将这些诱导得到的细胞移植回小鼠脑部海马区的齿状回后，iNSCs 可以正常存活并与周围神经元形成突触连接。上述实验证明 iNSCs 将会成为一种用于临床治疗神经退行性疾病和新药物筛选的细胞资源（Sheng et al., 2012）。该论文入选"2012 年度中国百篇最具影响力国际论文"，排名第 13 位。

研究发现，在神经前体细胞（neural progenitor cell，NPC）中过表达去甲基化酶 TET2 的催化结构域，可促进 NPC 向星形胶质细胞分化，抑制 NPC 向神经元分化，并且转录因子 OLIGO2 可直接结合到 TET2 基因的启动子区，调控其表达。在胚胎神经发育早期，转录因子 Ngn1 可与大脑中高表达的非编码 RNA(miR-9)的启动子结合，抑制 Jak-Stat 信号通路的活化，从而抑制星形胶质细胞的发生（Zhao et al., 2015a）。

3.3.3 发现疾病相关的重要表观遗传修饰

定位于 X 染色体上的 *MECP2* 基因突变是瑞特综合征的病因。之前的小鼠模型与临床患者存在较大差异，难以利用这些模型推动致病机理的研究，以及新治疗方法的研发。非人灵长类（如猕猴和食蟹猴）与人类具有高度相似的遗传背景及大脑结构，是开展脑发育及神经系统疾病最理想的实验动物。我国科学家于 2014 年初最早报道了利用靶向基因编辑方法 TALEN 成功编辑 *Mecp2* 基因的猴模型研究，相关结果发表在 *Cell Stem Cell* 上（Liu et al., 2014）。进一步研究证实，不同于啮齿类模型，*Mecp2* 基因突变的小猴表现出非常类似临床瑞特综合征（rett syndrome, RTT）患者的一系列病理和行为学特征，较啮齿类有无法比拟的优势，对未来更深入地开展 RTT 发病机理研究和治疗将产生深刻影响（Chen et al., 2017）。

科学家发现淋巴细胞特异性转录因子 Aiolos 通过改变 *p66Shc* 基因染

色质高级结构阻止增强子与启动子之间的相互作用，抑制 p66Shc 基因转录，促进肿瘤细胞逃逸失巢凋亡发生远端转移；提出实体肿瘤细胞通过"co-opt"淋巴细胞转录调控因子，模仿淋巴细胞的低黏附性和失巢凋亡抵抗特性，实现远端转移的新观点；证实染色质高级结构这一表观遗传调控在肿瘤发生中的重要作用（Li et al., 2014b）。这一工作发表后成为当期的"Featured article"，失巢凋亡领域的奠基人 Steven Frisch 同期对此发表了专题评论，认为"这一发现创新性地提出了肿瘤细胞利用淋巴细胞转录因子获得失巢凋亡抵抗特性"。美国 ISMMS 的 Powell 教授在 *American Journal of Respiratory and Critical Care Medicine* 上发表的"Update in Lung Cancer in 2014"一文中称"关于肿瘤细胞利用造血细胞信号机制的发现为肺癌发生机制研究开辟了一个新的有前途的方向"。该成果获得国家发明专利 2 项，其中 1 项于 2015 年转让给北京百奥赛图生物技术有限公司进行成果转化。

在 DNA 甲基转移酶（DNMT）家族中，*DNMT3A* 和 *DNMT3B* 行使从头甲基化的修饰作用。*DNMT3A* 的突变常见于多种血液系统恶性肿瘤，相反，*DNMT3B* 在血液系统中的突变鲜见报道。我国科学家利用 MLL-AF9 诱导的小鼠 AML 模型研究 *DNMT3B* 在急性髓系白血病中发挥的作用。通过增加细胞中白血病干细胞（leukemia stem cells, LSCs）的数量和促进细胞周期的活跃，*DNMT3B* 的缺失加速 MLL-AF9 白血病进程。*DNMT3B* 的缺失能够与 *DNMT3A* 缺失协同促进白血病发展。本研究对 DNA 甲基化转移酶在白血病发生中的作用提供了新的见解（Zhang et al., 2016）。

增强子是一类控制基因表达的重要调控元件，我国科学家发现由 H3K4me3 和 H3K27Ac 所共同标记的增强子可处于"过度活化态"，同时发现两个潜在的肿瘤抑制因子——RACK7/KDM5C 复合体是该过度活化态的负调控因子。RACK7 是一个潜在的染色质阅读器，KDM5C 是一种组蛋白去甲基化酶，两者协同，通过控制活化增强子处 H3K4me1 与 H3K4me3 的动态，充当活化增强子的"刹车"。丧失这样的增强子监管机制会导致细胞行为改变，有可能促成肿瘤发生（Shen et al., 2016）。

我国科学家采用 TAB-Seq（5hmC）测序方法在单碱基水平探究了 5mC 和 5hmC 在 ccRCC 中的重编程模式和规律，结合临床数据，结果显示 5hmC 水平降低而非 5mC 水平降低是肾癌更敏感的预后表观标记物。在裸鼠实验中，5hmC 的回调抑制了肿瘤的生长，这说明了 5hmC 在肾癌中的重编程不是一种伴随现象，更有可能是一种驱动力，暗示 5hmC 可以作为肾癌预后监测的新型表观遗传标记物和靶向治疗的靶点（Chen et al., 2016b）。

发现蛋白质去泛素化酶 USP7 介导的组蛋白去甲基化酶 PHF8 的稳定性赋予了肿瘤细胞抵抗 DNA 损伤的能力，促进乳腺癌发生（Wang et al., 2016），该研究发表时入选 *JCI* 当期的 "Editor's Picks"。同时，PHF8 突变被发现在与 X 染色体相关的智力缺陷等疾病诊断中起着重要的作用（Yu et al., 2010）。

科学家发现，转录中介体耦连基础转录机器和 RNA 聚合酶 II 精细调控基因转录。已知中介体复合物调控基因转录起始和延伸过程，但其在组蛋白修饰酶和蛋白翻译后修饰中的作用未曾有过报道。这一工作揭示了中介体 Med23 亚基特异性调控 H2B 单泛素化（H2Bub）控制特异靶基因的转录水平，来影响骨骼肌分化和肺癌发生过程，这在细胞命运决定过程中发挥了重要功能。这项工作揭示了新的转录延伸调控机制及其在细胞命运决定中的新功能，其成果作为封面论文发表在国际知名学术期刊上，受到了同行的高度评价。

肿瘤抑制因子 FoxO1 是诱导细胞自噬的关键蛋白。细胞质内的 FoxO1 与组蛋白去乙酰化酶 SIRT2 结合而保持非活性状态，但在应激情况下，FoxO1 与 SIRT2 脱离，变成活化状态的乙酰化 FoxO1，该活化状态的 FoxO1 又特异结合自噬关键蛋白 ATG7，激发细胞自噬过程。动物实验和临床肿瘤患者标本研究均证实这种细胞质内 FoxO1 引起的自噬是 FoxO1 抗肿瘤的主要原因之一。该发现将表观遗传修饰的组蛋白去乙酰化酶与细胞自噬以及肿瘤抑制功能有机联系起来。该论文发表在 *Nature Cell Biology* 上（Zhao et al., 2010）。

一直以来人们认为端粒的调控和与细胞代谢相关的线粒体功能调控是两条独立的细胞衰老调控通路。TIN2 蛋白能被召集到端粒中，与多个端粒调控因子，比如 TPP1，相互作用。TPP1 通过与 TIN2 的 N 末端发生反应从而调控 TIN2 的定位。我国科学家研究发现端粒体蛋白 TIN2 不仅对端粒功能的实现具有重要的作用，还可以在线粒体中进行翻译后修饰，并且能调控线粒体氧化磷酸化过程。通过 RNAi 敲除减少 TIN2 的表达，就会抑制糖酵解的发生和氧自由基的生成，并增加细胞中 ATP 水平和耗氧量，而 TIN2 在端粒和线粒体间的定位则依赖于 TPP1 的调控。这些研究结果表明，在端粒蛋白和代谢调控之间存在直接的关联，是端粒蛋白继调控癌症和衰老作用之外的又一重要作用机制。研究成果以封面文章"Mitochondrial Localization of Telomeric Protein TIN2 Links Telomere Regulation to Metabolic Control"发表在 *Molecular Cell* 上。

颅面骨由头部神经脊细胞发育分化而成，涉及神经脊细胞的命运决定、迁移和分化等重要过程，其中任何环节的异常都会导致颅面畸形。BMP 配体表达在咽区内胚层来源的咽囊，而其拮抗因子 Noggin3 表达于咽囊旁侧的软骨前体细胞中。研究发现，miR-92a 在斑马鱼咽部表达，主要通过维持咽部骨形成蛋白 BMP 的信号强度而保障咽部软骨的正常发育。该研究不仅阐明了 miR-92a 在软骨发育中的重要作用，还进一步说明 BMP 信号在咽部软骨形成过程中必须受到严格的调控，其活性高于或低于生理水平，都会导致严重的发育缺陷（Ning et al., 2013）。这一发现被 *F1000* 推荐，认为这项研究发现了导致颅面畸形的新的分子机制，是一项值得写入教科书的研究工作。

3.4 构建表观遗传学图谱，揭示胚胎发育表观遗传修饰特点和规律

采集高通量数据构建表观遗传学图谱，揭示胚胎发育早期、不同物种中表观遗传修饰的特点和遗传规律，丰富了我们对表观遗传信息网络的起源与进化的认识。

3.4.1 发现植入前胚胎中全基因组水平组蛋白修饰分布及变化规律

表观遗传重塑对受精过程中高度特化配子的重编程和胚胎发育十分重要。这些表观修饰的变化是胚胎基因组激活及第一次细胞谱系分化的关键，组蛋白的转录后修饰直接调控基因表达的激活或沉默。早期的研究中，利用抗体免疫荧光染色的方法研究发现，大部分组蛋白修饰在植入前胚胎的发育过程中都发生了明显的变化，而一些调节组蛋白修饰的酶的异常表达或缺失会导致胚胎发育异常甚至植入前胚胎的死亡。这些研究证明，组蛋白修饰的变化在早期胚胎发育的过程中起着很重要的作用。但是在植入前胚胎中，这些组蛋白修饰在基因组上是如何分布及变化的，以及这些变化是如何调控胚胎基因的表达和第一次细胞命运的分化的，还未知。通过改进最新的适用于低起始量细胞的 ULI-NChIP（ultra-low-input micrococcal nuclease-based native ChIP）技术，用极少量的细胞检测了小鼠植入前胚胎发育各个时期的组蛋白 H3K4me3 和 H3K27me3 修饰的变化情况，这两个修饰分别对应基因的激活和沉默。这是目前已知的第一次系统地对小鼠植入前胚胎的组蛋白修饰进行的全基因组水平上的检测。从全基因组水平上揭示了小鼠植入前胚胎发育过程中的组蛋白 H3K4me3 和 HK27me3 修饰的建立过程，并发现了宽的（broad）H3K4me3 修饰在植入前胚胎发育过程中对基因表达调控发挥了重要作用（Liu et al., 2016）。非常值得一提的是，

在本研究同一期 *Nature* 发表时，由本项目资助的另一篇发表在同期 *Nature* 上的"背靠背"的研究论文，发现基因组激活前（2 细胞前），不转录的卵子以及基因组激活前的 1 细胞、2 细胞早期，H3K4me3 呈现一种非经典形式（non-canonical H3K4me3），并且大量出现在非启动子区，比如基因间区（intergenic region）。通过在卵子中过表达去甲基化酶 KDM5B 来去除 H3K4me3，研究人员惊奇地发现这种非经典的 H3K4me3 对卵子的基因组沉默（而不是激活）是必需的（Zhang et al., 2016）。同期 *Nature* 发表评论"Developmental Biology: Panoramic Views of the Early Epigenome"（《发育生物学：早期表观基因组的全景展望》），指出这些研究共同展示了受精后及胚胎发育早期组蛋白修饰的变化细节，染色质开放程度对基因表达的调节，以及表观遗传信息是怎样在亲代和子代之间传递的。这些发现对研究胚胎发育异常、提高辅助生殖技术的成功率具有重要意义，将造福反复流产、胚胎停育、不孕不育患者。这也标志着我国在相关领域取得了具有世界影响力的研究成果。中国科协生命科学学会联合体发布的 2016 年度"中国生命科学领域十大进展"，本研究成果成功入选。

3.4.2　建立单细胞转录组测序及分析工具

单细胞 RNA 测序（single-cell RNA-sequencing，SCRS）是分析单个细胞或微量 RNA 中基因组表达的强有力的技术。与微阵列技术相比，SCRS 能检测出更多的转录组，灵敏度更高。它既能分析同一基因的多个转录本及其对应的蛋白类型，也能检测已知基因中新的剪接点，还具有准确度高、噪音低等优点。SCRS 有助于精准展示细胞编程、重编程分化过程中的细节变化。近年来研究人员开始利用这一技术来克服研究中起始样本量少的瓶颈。我国科学家通过采用自主研发的单细胞转录组分析技术，同时运用加权基因共表达网络分析（weighted gene co-expression network analysis，

WGCNA）技术，揭示了成年小鼠前脑神经发生区域 CD133+/GFAP− 室管膜（E）细胞的分子特征，发现了室管膜 CD133+/GFAP− 休眠细胞独特基因网络中的重要枢纽基因包含较多的免疫应答基因及血管生成因子受体编码基因。给予血管内皮生长因子（vascular endothelial growth factor, VEGF）可激活侧脑室以及第四脑室 CD133+ 室管膜神经干细胞（NSCs），加上碱性成纤维生长因子（bFGF）可诱导随后的神经谱系分化和迁移。这些研究结果表明了中枢神经系统中的整个脑室表面都存在休眠的室管膜神经干细胞，并揭示出了中枢神经系统损伤之后可让它们激活的丰富信号（Luo et al., 2015），通过结合膜片钳和单个神经元细胞转录组检测，揭示神经元成熟的分子标记及其与能量代谢的关系（Chen et al., 2016c），通过单细胞转录组分析揭示肝脏干细胞发生谱系和活化机制。

3.4.3 构建全基因组甲基化图谱，揭示表观遗传修饰的遗传与进化规律

在本重大研究计划支持下，我国科学家应用 Methy-Seq 技术测定低甲基化水平的物种——家蚕，构建了一张昆虫的单碱基分辨率甲基化谱——家蚕丝腺甲基化谱（Xiang et al., 2010）。首次在昆虫中建立了昆虫表观遗传学组，证实了昆虫表观遗传机制的存在和重要功能意义，澄清了长久以来人们对昆虫表观遗传系统的模糊认识，启发了后续研究者们重新思考昆虫 DNA 甲基化及其功能的研究。研究发表以来已被引用 170 多次。*Nature China* 将该成果作为亮点成果报道；*Nature Asia-Pacific* 网站将该成果作为研究亮点进行了题为"Threading Together a Map of Silkworm DNA Modifications"的报道；*Nature China* 网站对该成果发表题为"Epigenetics, Few and Far Between"评论文章，指出"这项工作表明可以从表观遗传学角度了解昆虫的进化，并且该研究结果为探讨表观遗传学对家蚕驯化的影

响提供了宝贵的数据"。该论文发表以来,已被三本英文专著 *Epigenetic Genetic Regulation and Epignomics*、*Insect Molecular Biology and Biochemistry* 和 *Honeybee Neurobiology and Behavior: A Tribute to Randolf Menzel* 引用。

  DNA 甲基化作为重要表观遗传机制调控基因的表达,影响一系列的生物学过程,如细胞命运决定、发育和组织、器官的稳态维持。DNA 甲基化以多种修饰方式,如 5mC、6mA(N6- 甲基腺嘌呤)和 4mC(N4- 甲基胞嘧啶),广泛存在于细菌、真核生物中。5mC 和其去甲基化过程中的衍生物 5hmC 在哺乳动物基因组 DNA 中被认为是碱基甲基化形式。与之不同的是,6mA 以较高丰度存在于原核生物及一些低等的真核生物,尤其是细菌中。6mA 修饰在 DNA 复制、修复、基因表达调控及宿主—病原体相互拮抗等方面发挥重要的作用。我国科学家研究证明了果蝇基因组中存在 6mA 修饰(Zhuang et al., 2015),并且证明该修饰在胚胎发育的早期阶段受到去甲基化酶 DMAD(果蝇 TET 同源蛋白)的精确调控,揭示了真核生物 DNA 新修饰形式,在表观遗传研究领域取得了原创性的突破。

  我国科学家用 MethylC-Seq 的方法测量了斑马鱼配子和早期胚胎单碱基分辨率全基因组图谱(见图 3.4),研究斑马鱼亲代 DNA 甲基化图谱遗传到子代中的规律(Jiang et al., 2013)。本研究发现父源的 DNA 稳定保持着精子的甲基化图谱,而母源 DNA 抛弃卵子甲基化图谱,并重编程为精子图谱,而这套图谱用于调控胚胎的早期发育。这一发现挑战了以往所认为的"早期胚胎发育调控的主要信息都储存在卵子中,而精子只提供另一套 DNA 序列而已"的传统观念。这一斑马鱼甲基化继承的研究证明:除了 DNA 序列被遗传外,表观遗传信息(DNA 甲基化图谱)也可以完整地遗传到子代中。研究成果发表时 *Cell* 以"Beyond DNA:Programming and Inheritance of Parental Methylomes"为题作专题评论。表观遗传信息可以遗传到子代,意味着在调控动物发育、表型,甚至疾病等方面,表观遗传信息的变异可能和遗传信息的变异一样起重要作用。研究发表之后,文章 1/4 的引用来自于进化领域，引发了关于表观遗传信息是否对进化起驱动作用的新思考。同时，在哺乳动物 DNA 甲基化重编程的研究中发现，在哺乳动物的早期发育中，甲基化的氧化产物在父源和母源基因组中都存在，因此无论父源还是母源 DNA 都存在主动去甲基化的方式，改写了长期以来关于哺乳动物早期胚胎发育过程中母源基因组通过被动稀释去甲基化的错误认识（Wang et al., 2014a）。

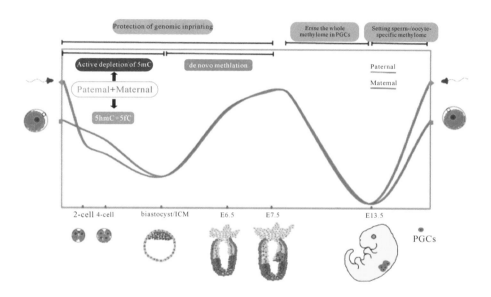

图 3.4　哺乳动物父源和母源的主动去甲基化方式

第4章 展 望

表观遗传学自20世纪80年代末期开始兴起。进入21世纪，随着技术手段的进步，这一新兴学科呈现出前所未有的快速发展态势。越来越多的生物物理学家、发育生物学家、化学家、生物信息学家与遗传学家一起投身于表观遗传学的探索当中。他们着眼国际学科前沿和发展动态，及时将相关学科的最新研究思想和技术手段应用到表观遗传学研究中，极大地提高了人们对表观遗传体系的全面认识。经过本重大研究计划的推动，我国学者在表观遗传学的分子基础及细胞重编程理论创新领域取得了长足的进步，为实现我国表观遗传学研究的全面跨越式发展做出了突出贡献。

当前，细胞重编程研究正处于快速发展的时期，很多方向上出现了原理和技术上的重大突破，热点层出不穷，变化日新月异，除了Yamanaka因子组合OSKM以外，科学家已经找到了更多的多能性重编程因子，如NANOG、PRDM14、SALL4、ESRRB、UTFL1、TET2、GLIS1等。为了防止这些外源性基因整合入目标细胞的基因组，提高系统的安全性，科学家在利用非病毒载体、RNA、可穿膜蛋白质以及化学小分子实现体细胞重编程和细胞转分化方面也取得了长足的进步。同时，细胞重编程过程中的表观遗传学研究进展非常迅速，最新的灵长类体细胞核移植的成功正是借助了表观遗传因子的使用。然而，现有的研究大多集中于关键因子的发现和功能验证，对重编程细胞中染色体动态、细胞表观遗传组学变化及其调控转录组变化、决定细胞命运转变的机制等，依然知之甚少。在这一大背景下，我国科学家应该抓住机遇，继续发挥学科交叉的优势，将在研究生命大分子结构和单细胞组学、基因编辑过程中所积累的技术、策略和经验应用到细胞重编程和表观遗传学研究中，紧盯生命科学的前沿问题，发展新技术、新方法，推动表观生物学研究从单分子走向多分子，从单一类型修饰走向多种类型修饰，从定性走向定量，从简单体系走向复杂体系，提高细胞重编程过程的效率和质量，在生命科学理论的发展上和技术革新上实现质的飞跃，为人类疾病的治疗带来美好的前景，开辟新的方向。

4.1　我国表观遗传研究待加强的方向

我国表观遗传领域需要加强以下几个薄弱方向的研究。

（1）加强表观遗传调控细胞分化、组织稳态以及器官发育与再生机理的研究，注重阐明表观遗传因子整合细胞内外环境和信号变化来调控细胞命运的分子机制。

（2）加强表观遗传调控机制在相关重大疾病中致病机理的研究，构建疾病相关表观遗传基因组图谱，加强与临床医学的交叉合作。

（3）加强发展新的模式生物，发挥我国物种多样性的优势，研究表观遗传因子对学习、记忆、社群动物行为以及获得性性状跨代遗传的机制，发现新的表观遗传现象，从物种进化的角度研究表观遗传因子对基因组稳定性和生物体性状的调控机理。

（4）加强细胞核染色质高级结构组成方式和动态变化的研究，进一步发展单细胞技术，并积极与物理学、材料化学以及计算生物学等学科交叉合作，阐明染色质高级结构组装的一般性原则和动态变化的调控机制。

4.2　我国表观遗传研究领域的战略需求

表观遗传调控在肿瘤、糖尿病、精神病及神经系统、生殖系统疾病等复杂疾病的发生发展中起着决定性的作用，而且生命个体对环境因素（包括营养、物理化学、心理因素等）的有序应答在很大程度上依赖于表观遗传调控网络的有效运行。表观遗传调控在植物发育、植物抗性、植物杂种优势的形成等方面也起着重要的作用。表观遗传学领域的基础生物学研究将帮助我们更清楚地了解相关复杂疾病的发病机制、个体对环境变化应答的方式和反应强度、植物发育和育种的调控机制等，为我们针对特定疾病开发新型治疗手段和原创性药物，改善个体生存状态，提高农业生产质量和粮食安全奠定基础。

通过本重大研究计划的实施，我国已经培养建立了涵盖表观遗传研究领域各个主要方向的是国际前沿水平的研究队伍，取得的诸多成果说明了研究队伍在表观遗传研究领域的实力。未来，我国有必要持续支持这些科研人员，保持我国该领域科学家队伍的稳定，这不但对我国保持该领域的国际地位至关重要，而且对支撑我国未来国计民生的发展至关重要。

4.3　深入研究的设想和建议

基于本重大研究计划已经取得的成果和总结，课题组提出以下研究设想。

（1）染色质生物学亟待和物理学、材料化学及计算生物学等学科的交叉，争取在方法学上取得突破，从而实现原位、实时、单细胞水平的染色质形态和高级结构的测定和跟踪观测，以及开展单细胞水平的表观遗传基因组研究。这对于表观遗传信息的建立、调控细胞命运及应答外界环境变化的研究具有重要意义。

（2）表观遗传学亟待深化与化学、药学、计算生物学等学科的交叉，筛选、鉴定、合成一批靶向表观遗传调控因子的小分子药物，并积极推进针对特定疾病的临床治疗研究，这些小分子药物在基础研究和临床上的应用具有重要的意义。

（3）表观遗传学亟待深化与数学、计算生物学等学科的交叉，对表观遗传基因组和表观遗传调控网络进行系统分析和模拟运算，并结合细胞内其他调控网络相互作用，从系统生物学的角度阐明表观遗传调控对个体发育和应答外界环境变化的重要意义。

（4）表观遗传学亟待深化与神经生物学、生理学、生物信息学等学科的交叉，从而发现新的表观遗传现象，探索新的表观遗传调控机制，拓宽现有的研究领域。

在本重大研究计划支持的重要研究方向的基础上，课题组建议加强对以下几个方向的关注和支持，继续提升我国表观遗传学研究在国际科学界的地位。

（1）继续加强细胞染色质高级结构的研究，特别是解析细胞不同分化状态、不同细胞类型的染色质高级结构组成方式和动态变化。我国在染色质高级结构方面，特别是核小体 30nm 高级结构方面，总体研究水平已经处于世界前列。30nm 染色质是以 4 个核小体为结构单元扭曲形成的，这种结构单元中的空隙刚好为表观遗传调控提供了一个窗口，为研究表观遗传调控的机理、解释表观遗传的基本问题提供了可能。加强这个方向的研究可以保持我国在该领域的优势。

（2）拓展对新发现的表观遗传修饰机制与功能的研究。我国科学家在发现新的表观调控因子方面取得了重要进展，例如新发现的 RNA m6A 修饰。下一步需要加强该方向的投入，同时深入研究新发现的表观遗传修饰的机制与功能，研究 DNA 表观修饰和 RNA 表观修饰之间可能的相互作用，加强表观遗传调控细胞分化、组织稳态以及器官发育与再生机理的研究。发展新的模式生物，从物种进化的角度研究表观遗传因子对基因组稳定性和生物体性状的调控机理。

（3）开发新技术，推出理论和新方法。发展和优化表观基因组编辑技术以及表观转录组单细胞表观遗传信息测定技术，拓展表观遗传网络的生物信息学算法，在单细胞水平上研究表观遗传组学和转录组学之间的网络互作。加强表观遗传调控机制在相关重大疾病中致病机理的研究，加强与临床医学的交叉合作。

（4）在单细胞水平上，重点研究OSKM及其他多能性因子在人源细胞重编程过程中对细胞表观遗传组学、转录组学的动态影响。提高细胞重编程的效率，提升重编程细胞的质量和重编程体系的稳定性、均一性，为基于细胞重编程的再生医学以及iPSC和转分化细胞的临床应用打下基础。

（5）加强对表观遗传对环境信号的响应及记忆研究，深入揭示其作用机制。生命机体内的表观遗传体系可以使拥有相同基因组的细胞在不同环境条件下呈现不同的表观基因组、转录组，从而分化出不同的形态功能。目前发现组蛋白修饰具有继承性，受到表观遗传修饰酶的反馈调节，这预示着生物体内存在表观遗传的响应和记忆体系。然而目前对其认识还不多。

参考文献

ANDREWS F H, STRAHL B D, KUTATELADZE T G, 2016. Insights into newly discovered marks and readers of epigenetic information. Nat Chem Biol, 12(9): 662-668.

CHEN J, CHEN X, LI M, et al, 2016a. Hierarchical Oct4 binding in concert with primed epigenetic rearrangements during somatic cell reprogramming. Cell Rep, 14(6): 1540-1554.

CHEN K, ZHANG J, GUO Z, et al, 2016b. Loss of 5-hydroxymethylcytosine is linked to gene body hypermethylation in kidney cancer. Cell Res, 26(1): 103-118.

CHEN P, ZHAO J, WANG Y, et al, 2013. H3. 3 actively marks enhancers and primes gene transcription via opening higher-ordered chromatin. Genes Dev, 27(19): 2109-2124.

CHEN T, HAO Y J, ZHANG Y, et al, 2015a. m(6)A RNA methylation is regulated by microRNAs and promotes reprogramming to pluripotency. Cell Stem Cell, 16(3): 289-301.

CHEN X, TANG S, ZHENG J S, et al, 2015b. Chemical synthesis of a two-photon-activatable chemokine and photon-guided lymphocyte migration in vivo. Nat Commun, 6: 7220.

CHEN X, ZHANG K, ZHOU L, et al, 2016c. Coupled electrophysiological recording and single cell transcriptome analyses revealed molecular mechanisms underlying neuronal maturation. Protein Cell, 7(3): 175-186.

CHEN Y, YU J, NIU Y, et al, 2017. Modeling rett syndrome using TALEN-Edited MECP2 mutant cynomolgus monkeys. Cell, 169(5): 945-955.

CHENG J, YANG H, FANG J, et al, 2015. Molecular mechanism for USP7-mediated DNMT1 stabilization by acetylation. Nat Commun, 6: 7023.

CRAMER P, 2014. A tale of chromatin and transcription in 100 structures. Cell, 159(5): 985-994.

DUTTA A, ABMAYR S M, WORKMAN J L, 2016. Diverse activities of histone acylations connect metabolism to chromatin function. Mol Cell, 63(4): 547-552.

FANG J, CHENG J, WANG J, et al, 2016. Hemi-methylated DNA opens a closed conformation of UHRF1 to facilitate its histone recognition. Nat Commun, 7: 11197.

FENG G, TONG M, XIA B, et al, 2016. Ubiquitously expressed genes participate in cell-specific functions via alternative promoter usage. EMBO Rep, 17(9): 1304-1313.

GAO J, ZHU Y, ZHOU W, et al, 2012. NAP1 family histone chaperones are required for somatic homologous recombination in Arabidopsis. Plant Cell, 24(4): 1437-1447.

GUO X, WANG L, LI J, et al, 2015. Structural insight into autoinhibition and histone H3-induced activation of DNMT3A. Nature, 517(7536): 640-644.

HU H, LIU Y, WANG M, et al, 2011. Structure of a CENP-A-histone H4 heterodimer in complex with chaperone hjurp. Genes Dev, 25(9): 901-906.

HU L, LU J, CHENG J, et al, 2015. Structural insight into substrate preference for TET-mediated oxidation. Nature, 527(7576): 118-122.

HUANG C, XIANG Y, WANG Y, et al, 2010. Dual-specificity histone demethylase kiaa1718 (kdm7a) regulates neural differentiation through FGF4. Cell Res, 20(2): 154-165.

HUANG P, HE Z, JI S, et al, 2011. Induction of functional hepatocyte-like cells from mouse fibroblasts by defined factors. Nature, 475(7356): 386-389.

JIA Y, LI P, FANG L, et al, 2016. Negative regulation of DNMT3A de novo DNA methylation by frequently overexpressed UHRF family proteins as a mechanism for widespread DNA hypomethylation in cancer. Cell Discov, 2: 16007.

JIANG L, ZHANG J, WANG J J, et al, 2013. Sperm, but not oocyte, DNA methylome is inherited by zebrafish early embryos. Cell, 153(4): 773-784.

LI L, SONG L, LIU C, et al, 2015. Ectodermal progenitors derived from epiblast stem cells by inhibition of Nodal signaling. J Mol Cell Biol, 7(5): 455-465.

LI W, CHEN P, YU J, et al, 2016a. FACT remodels the tetranucleosomal unit of chromatin fibers for gene transcription. Mol Cell, 64(1): 120-133.

LI W, SHUAI L, WAN H, et al, 2012a. Androgenetic haploid embryonic stem cells produce live transgenic mice. Nature, 490(7420): 407-411.

LI X, CUI XL, WANG J Q, et al, 2016b. Generation and application of mouse-rat allodiploid embryonic stem cells. Cell, 164(1-2): 279-292.

LI X, LIU L, YANG S, et al, 2014a. Histone demethylase KDM5B is a key regulator of genome stability. Proc Natl Acad Sci USA, 111(19): 7096-7101.

LI X, XU Z, DU W, et al, 2014b. Aiolos promotes anchorage independence by silencing p66Shc transcription in cancer cells. Cancer Cell, 25(5): 575-589.

LI Y, SABARI B R, PANCHENKO T, et al, 2016c. Molecular coupling of histone crotonylation and active transcription by AF9 YEATS domain. Mol Cell, 62(2): 181-193.

LI Z, FEI T, ZHANG J, et al, 2012b. BMP4 Signaling Acts via dual-specificity phosphatase 9 to control ERK activity in mouse embryonic stem cells. Cell Stem Cell, 10(2): 171-182.

LIN H, WANG Y, WANG Y, et al, 2010. Coordinated regulation of active and repressive histone methylations by a dual-specificity histone demethylase ceKDM7A from Caenorhabditis elegans. Cell Res, 20(8): 899-907.

LIU C P, XIONG C, WANG M, et al, 2012. Structure of the variant histone H3. 3-H4 heterodimer in complex with its chaperone DAXX. Nat Struct Mol Biol, 19(12): 1287-1292.

LIU H, CHEN Y, NIU Y, et al, 2014. TALEN-mediated gene mutagenesis in rhesus and cynomolgus monkeys. Cell Stem Cell, 14(3): 323-328.

LIU P, DOU X, LIU C, et al, 2015. Histone deacetylation promotes mouse neural induction by restricting nodal-dependent mesendoderm fate. Nat Commun, 6: 6830.

LIU X, GAO Q, LI P, et al, 2013. UHRF1 targets DNMT1 for DNA methylation through cooperative binding of hemi-methylated DNA and methylated H3K9. Nat Commun, 4: 1563.

LIU X, WANG C, LIU W, et al, 2016. Distinct features of H3K4me3 and H3K27me3 chromatin domains in pre-implantation embryos. Nature, 537(7621): 558-562.

LUO Y, COSKUN V, LIANG A, et al, 2015. Single-cell transcriptome analyses reveal signals to activate dormant neural stem cells. Cell, 161(5): 1175-1186.

NING G, LIU X, DAI M, et al, 2013. MicroRNA-92a upholds Bmp signaling by targeting noggin3 during pharyngeal cartilage formation. Dev Cell, 24(3): 283-295.

QIAO Y, WANG R, YANG X, et al, 2015a. Dual roles of histone H3 lysine 9 acetylation in human embryonic stem cell pluripotency and neural differentiation. J Biol Chem, 290(4): 2508-2520.

QIAO Y, WANG X, WANG R, et al, 2015b. AF9 promotes hESC neural differentiation through recruiting TET2 to neurodevelopmental gene loci for methylcytosine hydroxylation. Cell Discov, 1: 15017.

SHEN H, XU W, GUO R, et al, 2016. Suppression of enhancer overactivation by a RACK7-histone demethylase complex. Cell, 165(2): 331-342.

SHENG C, ZHENG Q, WU J, et al, 2012. Generation of dopaminergic neurons directly from mouse fibroblasts and fibroblast-derived neural progenitors. Cell Res, 22(4): 769-772.

SHI L, WANG J, HONG F, et al, 2011. Four amino acids guide the assembly or disassembly of Arabidopsis histone H3. 3-containing nucleosomes. Proc Natl Acad Sci USA, 108(26): 10574-10578.

SHI X L, GAO Y, YAN Y, et al, 2016. Improved survival of porcine acute liver failure by a bioartificial liver device implanted with induced human functional hepatocytes. Cell Res, 26(2): 206-216.

SHU J, ZHANG K, ZHANG M, et al, 2015. GATA family members as inducers for cellular reprogramming to pluripotency. Cell Res, 25(2): 169-180.

SHUAI L, WANG Y, DONG M, et al, 2015. Durable pluripotency and haploidy in epiblast stem cells derived from haploid embryonic stem cells in vitro. J Mol Cell Biol, 7(4): 326-337.

SONG F, CHEN P, SUN D, et al, 2014. Cryo-EM study of the chromatin fiber reveals a double helix twisted by tetranucleosomal units. Science, 344(6182): 376-380.

TAN M, LUO H, LEE S, et al, 2011. Identification of 67 histone marks and histone lysine crotonylation as a new type of histone modification. Cell, 146(6): 1016-1028.

VENKATESH S, WORKMAN J L, 2015. Histone exchange, chromatin structure and the regulation of transcription. Nat Rev Mol Cell Biol, 16(3): 178-189.

WANG L, ZHANG J, DUAN J, et al, 2014a. Programming and inheritance of parental DNA methylomes in mammals. Cell, 157(4): 979-991.

WANG Q, MA S, SONG N, et al, 2016. Stabilization of histone demethylase PHF8 by USP7 promotes breast carcinogenesis. J Clin Invest, 126(6): 2205-2220.

WANG T, SHA H, JI D, et al, 2014b. Polar body genome transfer for preventing the transmission of inherited mitochondrial diseases. Cell, 157(7): 1591-1604.

WU Y, LIANG D, WANG Y, et al, 2013. Correction of a genetic disease in mouse via use of crispr-cas9. Cell Stem Cell, 13(6): 659-662.

WU Y, ZHOU H, FAN X, et al, 2015. Correction of a genetic disease by crispr-cas9-mediated gene editing in mouse spermatogonial stem cells. Cell Res, 25(1): 67-79.

XIANG H, ZHU J, CHEN Q, et al, 2010. Single base-resolution methylome of the silkworm reveals a sparse epigenomic map. Nat Biotechnol, 28(5): 516-520.

XIONG J, ZHANG Z, CHEN J, et al, 2016a. Cooperative Action between SALL4A and TET Proteins in Stepwise Oxidation of 5-Methylcytosine. Mol Cell, 64(5): 913-925.

XIONG X, PANCHENKO T, YANG S, et al, 2016b. Selective recognition of histone crotonylation by double PHD fingers of MOZ and DPF2. Nat Chem Biol, 12(12): 1111-1118.

XU Z, SONG Z, LI G, et al, 2016. H2B ubiquitination regulates meiotic recombination by promoting chromatin relaxation. Nucleic Acids Res, 44(20): 9681-9697.

YANG Y, HU L, WANG P, et al, 2010. Structural insights into a dual-specificity histone demethylase ceKDM7A from Caenorhabditis elegans. Cell Res, 20(8): 886-898.

YU L, WANG Y, HUANG S, et al, 2010. Structural insights into a novel histone demethylase PHF8. Cell Res, 20(2): 166-173.

YU Z, ZHOU X, WANG W, et al, 2015. Dynamic phosphorylation of CENP-A at Ser68 orchestrates its cell-cycle-dependent deposition at centromeres. Dev Cell, 32(1): 68-81.

ZHANG B, ZHENG H, HUANG B, et al, 2016. Allelic reprogramming of the histone modification H3K4me3 in early mammalian development. Nature, 537(7621): 553-557.

ZHANG G, HUANG H, LIU D, et al, 2015. N6-methyladenine DNA modification in drosophila. Cell, 161(4): 893-906.

ZHAO D, GUAN H, ZHAO S, et al, 2016a. YEATS2 is a selective histone crotonylation reader. Cell Res, 26(5): 629-632.

ZHAO J, LIN Q, KIM K J, et al, 2015a. Ngn1 inhibits astrogliogenesis through induction of miR-9 during neuronal fate specification. Elife, 4: e06885.

ZHAO Q, ZHANG J, CHEN R, et al, 2016b. Dissecting the precise role of H3K9 methylation in crosstalk with DNA maintenance methylation in mammals. Nat Commun, 7: 12464.

ZHAO Y, YANG J, LIAO W, et al, 2010. Cytosolic FoxO1 is essential for the induction of autophagy and tumour suppressor activity. Nat Cell Biol, 12(7): 665-675.

ZHAO Y, ZHAO T, GUAN J, et al, 2015b. A XEN-like state bridges somatic cells to pluripotency during chemical reprogramming. Cell, 163(7): 1678-1691.

ZHENG Y, ZHANG H, WANG Y, et al, 2016. Loss of Dnmt3b accelerates MLL-AF9 leukemia progression. Leukemia, 30(12) : 2373-2384.

ZHONG C, YIN Q, XIE Z, et al, 2015. CRISPR-Cas9-mediated genetic screening in mice with haploid embryonic stem cells carrying a guide RNA library. Cell Stem Cell, 17(2): 221-232.

ZHONG C, ZHANG M, YIN Q, et al, 2016. Generation of human haploid embryonic stem cells from parthenogenetic embryos obtained by microsurgical removal of male pronucleus. Cell Res, 26(6): 743-746.

ZHU Z, LI C, ZENG Y, et al, 2017. PHB Associates with the HIRA complex to control an epigenetic-metabolic circuit in human ESCs. Cell Stem Cell, 20(2): 274-289 e277.

ZHU Z, WANG Y, LI X, et al, 2010. PHF8 is a histone H3K9me2 demethylase regulating rRNA synthesis. Cell Res, 20(7): 794-801.

成果附录

附录1　重要论文目录

本重大研究计划执行期间，我国科学家在国际学术刊物发表 SCI 论文 800 余篇。其中包括 *Nature* 8 篇，*Science* 2 篇，*Cell* 13 篇，*Nature* 系列 39 篇，*Cell Stem Cell* 11 篇，*Cell Research* 46 篇，*PNAS* 25 篇。全部论文引用总计超过 1.8 万次，单篇引用最高 1200 余次（时间统计截至 2017 年 10 月）。这表明了我国科学家在表观遗传学领域的雄厚实力和国际领先地位。其中 20 篇代表性学术论文及被引用情况、已发表重要论文详见下文。

附录 1.1 代表性学术论文及被引用情况

1. HUANG P, HE Z, JI S, et al. Induction of functional hepatocyte-like cells from mouse fibroblasts by defined factors. Nature, 2011, 475(7356): 386-389. (Web of Science 引用次数 377, 高被引论文)
2. LI W, SHUAI L, WAN H, et al. Androgenetic haploid embryonic stem cells produce live transgenic mice. Nature, 2012, 490(7420): 407-411. (Web of Science 引用次数 69)
3. HU L, LU J, CHENG J, et al. Structural insight into substrate preference for TET-mediated oxidation. Nature, 2015, 527(7576): 118-122. (Web of Science 引用次数 35)
4. GUO X, WANG L, LI J, et al. Structural insight into autoinhibition and histone H3-induced activation of DNMT3A. Nature, 2015, 517(7536): 640-644. (Web of Science 引用次数 58)
5. ZHANG BJ, ZHENG H, HUANG B, et al. Allelic reprogramming of the histone modification H3K4me3 in early mammalian development. Nature, 2016, 537(7621): 553-557. (Web of Science 引用次数 22)
6. LIU X, WANG C, LIU W, et al. Distinct features of H3K4me3 and H3K27me3 chromatin domains in pre-implantation embryos. Nature, 2016, 537(7621): 558-562. (Web of Science 引用次数 22)
7. SONG F, CHEN P, SUN DP, et al. Cryo-EM study of the chromatin fiber reveals a double helix twisted by tetranucleosomal units. Science, 2014, 344(6182): 376-380. (Web of Science 引用次数 134, 高被引论文)
8. HOU P, LI Y, ZHANG X, et al. Pluripotent stem cells induced from mouse somatic cells by small-molecule compounds. Science, 2013, 341(6146): 651-654. (Web of Science 引用次数 465, 高被引论文)
9. JIANG L, ZHANG J, WANG J J, et al. Sperm, but not oocyte, DNA methylome is inherited by zebrafish early embryos. Cell, 2013, 153(4): 773-784. (Web of Science 引用次数 131, 高被引论文)
10. SHU J, WU C, WU Y, et al. Induction of pluripotency in mouse somatic cells with lineage specifiers. Cell, 2013, 153(5): 963-975. (Web of Science 引用次数 129)
11. WANG L, ZHANG J, DUAN J L, et al. Programming and inheritance of parental DNA methylomes in mammals. Cell, 2014, 157(4): 979-991. (Web of Science 引用次数 123, 高被引论文)
12. ZHANG GQ, HUANG H, LIU D, et al. N 6-methyladenine DNA modification in Drosophila. Cell, 2015, 161(4): 893-906. (Web of Science 引用次数 76, 高被引论文)
13. ZHAO Y, ZHAO T, GUAN J, et al. A XEN-like state bridges somatic cells to pluripotency during chemical reprogramming. Cell, 2015, 163(7): 1678-1691. (Web of Science 引用次数 31)

14. LUO Y, COSKUN V, LIANG A, et al. Single-cell transcriptome analyses reveal signals to activate dormant neural stem cells. Cell, 2015, 161(5): 1175-1186. (Web of Science 引用次数 46)
15. SHEN H, XU W, GUO R, et al. Suppression of enhancer overactivation by a RACK7-histone demethylase complex. Cell, 2016, 165(2): 331-342. (Web of Science 引用次数 16)
16. LI X, CUI XL, WANG J Q, et al. Generation and application of mouse-rat allodiploid embryonic stem cells. Cell, 2016, 164(1): 279-292. (Web of Science 引用次数 5)
17. LI W, TENG F, LI T, et al. Simultaneous generation and germline transmission of multiple gene mutations in rat using CRISPR-Cas systems. Nat biotechnol, 2013, 31(8): 684-686. (Web of Science 引用次数 184)
18. LI X, ZUO X, JING J, et al. Small-molecule-driven direct reprogramming of mouse fibroblasts into functional neurons. Cell Stem Cell, 2015, 17(2): 195-203. (Web of Science 引用次数 71, 高被引论文)
19. WU Y, LIANG D, WANG Y, et al. Correction of a genetic disease in mouse via use of CRISPR-Cas9. Cell Stem Cell, 2013, 13(6): 659-662. (Web of Science 引用次数 160)
20. WANG Y, HE L, DU Y, et al. The long noncoding RNA lncTCF7 promotes self-renewal of human liver cancer stem cells through activation of Wnt signaling. Cell Stem Cell, 2015, 16(4): 413-425. (Web of Science 引用次数 74, 高被引论文)

附录1.2 已发表的重要论文

1. XIA M F, BIAN M J, YU Q L, et al. Cold water stress attenuates dopaminergic neurotoxicity induced by 1-methyl-4-phenyl-1, 2, 3, 6-tetrahydro yridine in mice. Acta Biochim Biophys Sin, 2011, 43: 448-454.

2. HU S B, CHENG L, WEN B. Large chromatin domains in pluripotent and differentiated cells. Acta Biochim Biophys Sin, 2012, 44: 48-53.

3. CHEN Y C, ZHU W G. Biological function and regulation of histone and non-histone lysine methylation in response to DNA damage. Acta Biochim Biophys Sin, 2016, 48(7): 603-616.

4. ZHANG S, ZHOU X, LIU S, et al. MYH9-related disease: description of a large chinese pedigree and a survey of reported mutations. Acta Haematologica, 2014, 132(2): 193-198.

5. ZHANG R, HAN M, ZHENG B, et al. Krüppel-like factor 4 interacts with p300 to activate mitofusin 2 gene expression induced by all-trans retinoic acid in vsmcs. Acta Pharmacol Sin, 2010, 31: 1293-1302.

6. Zhang P, Cai L, Yang XB, et al. Sustained delivery growth factors with polyethyleneimine-modified nanoparticles promote embryonic stem cells differentiation and liver regeneration. Adv Sci (Weinh), 2016, 3(8): 1500393.

7. HAN Y X, HAN D L, YAN Z, et al. Stress-associated H3K4 methylation accumulates during postnatal development and aging of rhesus macaque brain. Aging Cell, 2012, 11(6): 1055-1064.

8. BO X, SHEN J, CHEN W Y, et al. Wormfarm: a quantitative control and measurement device toward automated caenorhabditis elegans aging analysis. Aging Cell, 2013, 12(3): 398-409.

9. HAN X, LIU D, ZHANG Y, et al. Akt regulates TPP1 homodimerization and telomere protection. Aging Cell 2013, 12(6): 1091-1099.

10. LUO Z, FENG X, WANG H, et al. Mir-23a induces telomere dysfunction and cellular senescence by inhibiting TRF2 expression. Aging Cell, 2015, 14(3): 391-399.

11. WANG D, HOU L, SHU H N, et al. LIN-28 balances longevity and germline stem cell number in Caenorhabditis elegans through let-7/AKT/DAF-16 axis. Aging Cell, 2017, 16(1): 113-124.

12. ZHU L L, LIU Y, CUI A F, et al. PGC-1 coactivates estrogen-related receptor-to induce the expression of glucokinase. Am J Physiol Endocrinol Metab, 2010, 298: E1210-E1218.

13. ZHANG J, ZHANG Q, DARRIN L, et al. A bayesian method for disentangling

dependent structure of epistatic interaction. Am J Biostat, 2011, 2(1): 1-10.

14. WU X Q, CAO Y S, JUNWEI N, et al. Genetic and pharmacological inhibition of rheb1-mTORC1 signaling exerts cardioprotection against adverse cardiac remodeling in mice. Am J Biostat, 2013, 182: 2005-2014.

15. YUAN B Y, WAN P, CHU D D, et al. A cardiomyocyte-specific wdr1 knockout demonstrates essential functional roles for actin disassembly during myocardial growth and maintenance in mice. Am J Biostat, 2014, 184: 1967-1980.

16. WU J, AN Y, PU H, et al. Enrichment of serum low-molecular-weight proteins using C18 absorbent under urea/dithiothreitol denatured environment. Anal Biochem, 2010, 398(1): 34-44.

17. MAO Y, LIN J, ZHOU A, et al. Quikgene: a gene synthesis method integrated with ligation free cloning. Anal Biochem, 2011, 415(1): 21-26.

18. WANG H, ZHOU N, DING F, et al. An efficient approach for site-directed mutagenesis using central overlapping primers. Anal Biochem, 2012, 418(1): 304-306.

19. LIU Y, ZHUANG D, HOU R, et al. Shotgun proteomic analysis of microdissected postmortem human pituitary using complementary two-dimensional liquid chromatography coupled with tandem mass spectrometer. Anal Chim Acta, 2011, 688: 183-190.

20. ZHANG K, ZHU Y, HE X, et al. Systematic screening of protein modifications in four kinases using affinity enrichment and mass spectrometry analysis with unrestrictive sequence alignment. Anal Chim Acta, 2011, 691(1-2): 62-67.

21. CAI T X, SHU Q B, HOU J J, et al. Rapid profiling and relative quantitation of phosphoinositides by multiple precursor ion scanning based on phosphate methylation and isotopic labeling. Anal Chem, 2015, 87(1): 513-521.

22. ZHANG K, TIAN S S, FAN E G. Protein lysine acetylation analysis: current MS-based proteomic technologies. Analyst, 2013, 138: 1628-1636.

23. ZHANG G, JING X, WANG X, et al. Contribution of the proinflammatory cytokine IL-18 in the formation of human nasal polyps. Anat Rec (Hoboken), 2011, 294: 953-960.

24. ZHANG G, SHAO J, SU C, et al. Distribution change of mast cells in human nasal polyps. Anat Rec (Hoboken), 2012, 295: 758-763.

25. YIN J, DING J, HUANG L, et al. SND1 affects proliferation of hepatocellular carcinoma cell line SMMC-7721 by regulating IGFBP3 expression. Anat Rec (Hoboken), 2013, 296: 1568-1575.

26. WANG X, LIU X, FANG J, et al. Coactivator p100 protein enhances STAT6-dependnent transcriptional activation but has no effect on STAT1-mediated gene

transcription. Anat Rec, 2010, 293: 1010-1016.

27. YANG J, ZHENG Z, YAN X, et al. Integration of autophagy and anoikis resistance in solid tumors. Anat Rec, 2013, 296(10): 1501-1508.

28. CHEN S M, CHEN X M, LU Y L, et al. Cofilin is correlated with sperm quality and influences sperm fertilizing capacity in humans. Andrology, 2016, 4(6): 1064-1072.

29. YOKO S, HU J J, MICHAEL M K, et al. Mechanisms shaping the membranes of cellular organelles. Annu Rev Cell Dev Biol, 2009, 25: 329-354.

30. ZHANG Y X, DU Y Z, LE W D, et al. Redox control of the survival of healthy and diseased cells. Antioxid Redox Signal, 2011, 15: 2867-2908.

31. DU Y, XIA Y, PAN X, et al. Fenretinide targets chronic myeloid leukemia stem/progenitor cells by regulation of redox signaling. Antioxid Redox Signal, 2014, 20(12): 1866-1880.

32. ZHANG X H, ZHENG B, GU C, et al. TGF-β1 downregulates AT1 receptor expression via PKC-Mediated Sp1 dissociation From KLF4 and smad-mediated PPAR- association with KLF4. Arterioscler Thromb Vasc Biol, 2012, 32: 1015-1023.

33. YANG J, ZHAO Y, MA K, et al. Deficiency of hepatocystin induces autophagy through an mTOR-dependent pathway. Autophagy, 2011, 7(7): 748-759.

34. ZHOU J, LIAO W, YANG J, et al. FOXO3 induces FOXO1-dependent autophagy by activating the AKT1 signaling pathway. Autophagy, 2012, 8(12): 1712-1723.

35. WEI F Z, CAO Z, WANG X, et al. Epigenetic regulation of autophagy by the methyltransferase EZH2 through an MTOR-dependent pathway. Autophagy, 2015, 11(12): 2309-2322.

36. SHANG Y, WANG H, JIA P, et al. Autophagy regulates spermatid differentiation via degradation of PDLIM1. Autophagy, 2016, 12(9): 1575.

37. MA K, FU W, TANG M, et al. PTK2-mediated degradation of ATG3 impedes cancer cells susceptible to DNA damage treatment. Autophagy, 2017, 13(3): 579-591.

38. MI Y, ZHANG Y, SHEN Y F. Mechanism of jmjC-containing protein hairless in the regulation of vitamin D receptor function. BBA-Mol Basis Dis, 2011, 1812: 1675-1680.

39. LIN X, WANG Q, CHENG Y, et al. Changes of protein expression profiles in the amygdala during the process of morphine-induced conditioned place preference in rats. Behav Brain Res, 2011, 221(1): 197-206.

40. ZHANG C G, JIA Z Q, LI B H, et al. B-catenin/TCF/LEF1 can directly regulate phenylephrine-induced cell hypertrophy and Anf transcription in cardiomyocytes. Biochem Biophys Res Commun, 2009, 390(2): 258-262.

41. WANG H Q, XU Y X, ZHAO X Y, et al. Overexpression of F(0)F(1)-ATP synthase

alpha suppresses mutant huntingtin aggregation and toxicity in vitro. Biochem Biophys Res Commun, 2009, 390: 1294-1298.
42. MENG Q Z, JIA Z Q, WANG W P, et al. Mapping of the minimal internal ribosome entry site element in the human embryonic stem cell gene OCT4B mRNA. Biochem Biophys Res Commun, 2010, 394: 750-754.
43. MENG Q Z, JIA Z Q, WANG W P, et al. Inhibitor of DNA binding 1 (Id1) induces differentiation and proliferation of mouse embryonic carcinoma P19CL6 cells. Biochem Biophys Res Commun, 2011, 412(2): 253-259.
44. JIANG L L, PAN X L, CHEN Y, et al. Preferential involvement of both ROS and ceramide in fenretinide-induced apoptosis of HL60 rather than NB4 and U937 cells. Biochem Biophys Res Commun, 2011, 405: 314-318.
45. CHEN W, JIA W, WANG K, et al. Retinoic acid regulates germ cell differentiation in mouse embryonic stem cells through a Smad-dependent pathway. Biochem Biophys Res Commun, 2012, 418(3): 571-577.
46. DU F, ZHANG M, LI X, et al. Dimer monomer transition and dimer re-formation play important role for ATM cellular function during DNA repair. Biochem Biophys Res Commun, 2014, 452(4): 1034-1039.
47. YAO L, LI Y, DU F, et al. Histone H4 Lys 20 methyltransferase SET8 promotes androgen receptor-mediated transcription activation in prostate cancer. Biochem Biophys Res Commun, 2014, 450(1): 692-696.
48. ZHU L, KE Y, SHAO D, et al. PPAR γ co-activator-1 α co-activates steroidogenic factor 1 to stimulate the synthesis of luteinizing hormone and aldosterone. Biochem J, 2010, 432: 473-483.
49. WANG R, KONG X, CUI A, et al. Sterol-regulatory-element-binding protein 1c mediates the effect of insulinon the expression of Cidea in mouse hepatocytes. Biochem J, 2010, 430: 245-254.
50. DI R M, WU X Q, CHANG Z, et al. S6K inhibition renders cardiac protection against myocardial infarction through PDK1 phosphorylation of Akt. Biochem J, 2012, 441: 199-207.
51. WEI J, WANG J, ZHANG W, et al. pVHL acts as a downstream target of E2F1 to suppress E2F1 activity. Biochem J, 2014, 457: 185-195.
52. YANG Y, WANG L L, LI Y H, et al. Effect of CpG island methylation on microRNA expression in the k-562 cell line. Biochemical genetics, 2012, 50: 122-134.
53. MA M Y, ZHOU L, GUO X J, et al. Decreased cofilin1 expression is important for compaction during early mouse embryo development. Biochim Biophys Acta-MCR, 2009, 1793(12): 1804-1810.

54. WANG X L, FAN J J, ZHU Y, et al. Histone H3k4 Methyltransferase Mll1 Regulates Protein Glycosylation and Tunicamycin-Induced Apoptosis through Transcriptional Regulation. Biochim Biophys Acta-MCR, 2014, 1843: 2592-2602.
55. ZHU Y, DONG A, SHEN W H. Histone variants and chromatin assembly in plant abiotic stress responses. Biochimica et Biophysica Acta-Gene Regulatory Mechanisms, 2012, 1819: 343-348.
56. HUANG J L, LIU Y, ZHANG W, et al. EResponse net: a package prioritizing candidate disease genes through cellular pathways. Bioinformatics, 2011, 27(16): 2319-2320.
57. HE C, MICHAEL Q Z, WANG X W. MICC: an R package for identifying chromatin interactions from ChIA-PET data. Bioinformatics, 2015, 31(23): 3832-3834.
58. YANG S, HARI K Y, LI X R, et al. Correlated evolution of transcription factors and their binding sites. Bioinformatics, 2011, 27(21): 2972-2978.
59. ROGE X, ZHANG X G. RNAseq viewer: visualization tool for RNA-Seq data. Bioinformatics, 2014, 30(6): 891-892.
60. WU H, ZHANG Z, HU S N, et al. On the molecular mechanism of GC content variation among eubacterial genomes. Biology Direct, 2012, 7: 2.
61. BI Y, LV Z, WANG Y H, et al. A key epigenetics-related factor, plays a crucial role in normal early embryonic development in mice. Biol Reprod, 2011, 84(4): 756-764.
62. ZHANG H, MA Y, GU J, et al. Reprogramming of somatic cells via TAT-mediated protein transduction of recombinant factors. Biomaterials, 2012, 33: 5047-5055.
63. GE W S, LIU Y S, CHEN T, et al. The epigenetic promotion of osteogenic differentiation of human adipose-derived stem cells by the genetic and chemical blockade of histone demethylase LSD1. Biomaterials, 2014, 35: 6015-6025.
64. HUANG H J, ZHANG X, HU X Q, et al. A functional biphasic biomaterial homing mesenchymal stem cells for in vivo cartilage regeneration. Biomaterials, 2014, 35(2014): 9608-9619.
65. WANG J Y, CHEN F L, LIU L W, et al. Engineering EMT using 3D micro-scaffold to promote hepatic functions for drug hepatotoxicity evaluation. Biomaterials, 2016, 91: 11-22.
66. GAO R, XIU W, ZHANG L, et al. Direct induction of neural progenitor cells transiently passes through a partially reprogrammed state. Biomaterials, 2017, 119: 53-67.
67. HAO P, DUAN H M, HAO F, et al. Neural repair by NT3-chitosan via enhancement of endogenous neurogenesis after adult focal aspiration brain injury. Biomaterials, 2017, 140: 88-102.

68. XI J, XU S, WU L, et al. Design, synthesis and biological activity of 3-oxoamino-benzenesulfonamides as selective and reversible LSD1 inhibitors. Bioorg Chem, 2017, 72: 182-189.

69. HAN J D. An aging program at the systems level? Birth defects research. Part C, Embryo today: reviews, 2012, 96(2): 206-211.

70. WANG L, LIU T H, XU L J, et al. Fev regulates hematopoietic stem cell development via ERK signaling. Blood, 2013, 122: 367-375.

71. LI Y, GAO L, LUO X, et al. Epigenetic silencing of microRNA-193a contributes to leukemogenesis in t(8;21) acute myeloid leukemia by activating the PTEN/PI3K signal pathway. Blood, 2013, 121: 499-509.

72. ZHANG H, CHENG H, WANG Y, et al. Reprogramming of Notch1-induced acute lymphoblastic leukemia cells into pluripotent stem cells in mice. Blood Cancer J, 2016, 6: e444.

73. CHEN S, ZENG M, SUN H Q, et al. Zebrafish dnd protein binds to 3'UTR of geminin mRNA and regulates its expression. BMB Rep, 2010, 43(6): 438-444.

74. LU Y L, LI C, ZHANG K, et al. Identification of piRNAs in Hela cells by massive parallel sequencing. BMB Rep, 2010, 43(9): 635-641.

75. CHEN H, SUN H, TAO D, et al. Znf451 affects primitive hematopoiesis by regulating transforming growth factor-β signaling. BMB Rep, 2014, 47(1): 21-26.

76. WU C, ZHU J, ZHANG X G. Network-based differential gene expression analysis suggests cell cycle related genes regulated by E2F1 underlie the molecular difference between smoker and non-smoker lung adenocarcinoma. BMC Bioinformatics, 2013, 14: 365.

77. TAO Y, ZHENG W, JIANG Y, et al. Nucleosome organizations in induced pluripotent stem cells reprogrammed from somatic cells belonging to three different germ layers. BMC Biol, 2015, 21(12): 109.

78. ZHAN Y, WEI Y, CHEN P, et al. Expression, purification and biological characterization of the extracellular domain of CD40 from Pichia pastoris. BMC Biotechnol, 2016, 16: 8.

79. WANG G Z, MARINI S, MA X Y, et al. Improvement of dscam homophilic binding affinity throughout drosophila evolution. BMC Evol Biol, 2014, 14: 186.

80. ZHANG P, NI X J, GUO Y, et al. Proteomic-based identification of maternal proteins in mature mouseoocytes. BMC Genomics, 2009, 10: 348.

81. LI T, WU R, ZHANG Y, et al. A systematic analysis of the skeletal muscle miRNA transcriptome of chicken varieties with divergent skeletal muscle growth identifies novel miRNAs and differentially expressed miRNAs. BMC Genomics, 2011, 12(1):

186.

82. XIANG H, LI X, DAI F, et al. Comparative methylomics between domesticated and wild silkworms implies possible epigenetic influences on silkworm domestication. BMC Genomics, 2013, 14: 646-651.
83. SLEUMER M C, WEI G, WANG Y, et al. Regulatory elements of Caenorhabditis elegans ribosomal protein genes. BMC Genomics, 2012, 13: 433.
84. GUO W L, PETKO F, YAN W H, et al. BS-Seeker2: a versatile aligning pipeline for bisulfite sequencing data. BMC Genomics, 2013, 14: 774.
85. BAO S Y, ZHOU X Y, ZHANG L C, et al. Prioritizing genes responsible for host resistance to influenza using network approaches. BMC Genomics, 2013, 14: 816.
86. CUI H F, ZHANG X G. Alignment - free supervised classification of metagenomes by recursive SVM. BMC Genomics, 2013, 14: 641.
87. LIU Z, ZHANG P, TANG S, et al. Urine real-time polymerase chain reaction detection for children virus pneumonia with acute human cytomegalovirus infection. BMC Infect Dis, 2014, 14: 245.
88. LIU Z, ZHANG P, HE X, et al. New multiplex real-time PCR approach to detect gene mutations for spinal muscular atrophy. BMC Neurol, 2016, 16: 141.
89. ZHANG T W, LUO Y F, CHEN Y P, et al. BIGrat: a repeat resolver for pyrosequencing-based re-sequencing with newbler. BMC Res Notes, 2012, 5(1): 567.
90. SHI S, LIN J, CAI Y, et al. Dimeric structure of p300/CBP associated factor. BMC Struct Biol, 2014, 14: 1402.
91. FANG H, JIN W, YANG Y, et al. An organogenesis network-based comparative transcriptome analysis for understanding early human development in vivo and in vitro. BMC Syst Biol, 2011, 5: 108.
92. HU Z R, QIAN M P, MICHAEL Q Z. Novel Markov model of induced pluripotency predicts gene expression changes in reprogramming. BMC Syst Biol, 2011, 5(S2): 8-19.
93. YAN L, ZHAO H Y, ZHANG Y, et al. Differential effects of AdOx on gene expression in P19 embryonal carcinoma cells. BMC Neurosci, 2012, 13: 6.
94. TANG W, ZHU Y, GAO J, et al. MicroRNA-29a promotes colorectal cancer metastasis by regulating matrix metalloproteinase 2 and E-cadherin via KLF4. Br J Cancer, 2014, 110(2): 450-458.
95. YAN J, SUN X B, WANG H Q. Chronic restraint stress alters the expression and distribution of phosphorylated tau and MAP2 in cortex and hippocampus of rat brain. Brain Research, 2010, 1347: 132-141.
96. WANG H Q, SUN X B, XU Y X. Astaxanthin upregulates heme oxygenase-1

expression through ERK1/2 pathway and its protective effect against beta-amyloid-induced cytotoxicity in SH-SY5Y cells. Brain Research, 2010, 1360: 159-167.
97. LI T, SONG B, SONG B, et al. Systematic identification of class I HDAC substrates. Brief Bioinform, 2014, 15(6): 963-972.
98. QIN R, LI K, QI X, et al. β-arrestin1 promotes the progression of chronic myeloid leukemia by regulating BCR/ABL H4 acetylation. Brit J Cancer, 2014, 111: 568-576.
99. TANG L Q, CHEN Q Y, GUO S S, et al. The impact of plasma epstein–barr virus DNA and fibrinogen on nasopharyngeal carcinoma prognosis: an observational study. Brit J Cancer, 2014, 111: 1102-1111.
100. HONG Z, LI H, LI L, et al. Different expression patterns of histone H3K27 demethylases in renal cell carcinoma and bladder cancer. Cancer Biomark, 2016, 18(2): 125-131.
101. DUAN C W, SHI J, CHEN J, et al. Leukemia propagating cells rebuild an evolving niche in response to therapy. Cancer Cell, 2014, 25: 778-793.
102. LI X, XU Z, DU W, et al. Aiolos promotes anchorage independence by silencing p66shc transcription in cancer cells. Cancer Cell, 2014, 25(5): 575-589.
103. SI W Z, HUANG W, ZHENG Y, et al. Dysfunction of the reciprocal feedback loop between GATA3- and ZEB2-nucleated repression programs contributes to breast cancer metastasis. Cancer Cell, 2015, 27: 822-836.
104. LIN S, ZHOU X, LIU X H, et al. FOXK2 Elicits Massive Transcription Repression and Suppresses the Hypoxic Response and Breast Cancer Carcinogenesis. Cancer Cell, 2016, 30: 708-722.
105. HU H L, YANG Y, JI Q H, et al. CRL4B catalyzes H2AK119 Monoubiquitination and coordinates with PRC2 to promote tumorigenesis. Cancer Cell, 2012, 22: 781-795.
106. ZHANG Y X, WANG C, WANG H W, et al. Combination of tetrandrine with cisplatin enhances cytotoxicity through growth suppression and apoptosis in ovarian cancer in vitro and in vivo. Cancer Lett, 2011, 304: 21-32.
107. DU W, JIANG Y, ZHENG Z, et al. Feedback loop between p66(Shc) and Nrf2 promotes lung cancer progression. Cancer Lett, 2013, 337(1): 58-65.
108. FENG H J, QIN Z Y, ZHANG X G. Opportunities and methods for studying alternative splicing in cancer with RNA-Seq. Cancer Lett, 2013, 340(2): 179-191.
109. JIANG S, ZHANG H W, LU M H, et al. MicroRNA-155 functions as an oncomir in breast cancer by targeting the suppressor of cytokine signaling 1 gene. Cancer Research, 2010, 70(8): 3119-3127.
110. WANG L, ZHANG L F, WU J, et al. IL-1b-Mediated repression of microRNA-101

is crucial for inflammation-promoted lung tumorigenesis. Cancer Research, 2014, 74(17): 4720-4730.

111. YU L, LIU X, CUI K, et al. SND1 acts downstream of tgfb1 and up stream of smurf1 to promote breast cancer metastasis. Cancer Research, 2015, 75: 1275-1286.

112. YUAN P, HE X H, RONG T F, et al. KRAS/NF-kB/YY1/miR-489 signaling axis controls pancreatic cancer metastasis. Cancer Research, 2017, 77 (1): 100-111.

113. ZHANG J, LIANG Q, LIU D, et al. SOX4 induces epithelial-mesenchymal transition and contributes to breast cancer progression. Cancer Research, 2012, 72(17): 4597-4608.

114. YU K, ZHENG B, HAN M, et al. ATRA activates and PDGF-BB represses the SM22 promoter through KLF4 binding to, or dissociating from, its cis-DNA elements. Cardiovasc Res, 2011, 90: 464-474.

115. YUE R, KANG J, ZHAO C, et al. Beta-arrestin1 regulates zebrafish hematopoiesis through binding to YY1 and relieving polycomb group repression. Cell, 2009, 139(3): 535-546.

116. JIANG L, ZHANG J, WANG J J, et al. Sperm, but not oocyte, DNA methylome is inherited by zebrafish early embryos. Cell, 2013, 153: 773-784.

117. SHU J, WU C, WU Y, et al. Induction of pluripotency in mouse somatic cells with lineage specifiers. Cell, 2013, 153(5): 963-975.

118. WANG L, ZHANG J, DUAN J L, et al. Programming and inheritance of parental DNA methylomes in mammals. Cell, 2014, 157: 979-991.

119. ZHANG G Q, HUANG H, LIU D, et al. N6-methyladenine DNA modification in drosophila. Cell, 2015, 161: 893-906.

120. ZHAO Y, ZHAO T, GUAN J, et al. A XEN-like state bridges somatic cells to pluripotency during chemical reprogramming. Cell, 2015, 163(7): 1678-1691.

121. LUO Y, COSKUN V, LIANG A, et al. Single-cell transcriptome analyses reveal signals to activate dormant neural stem cells. Cell, 2015, 161(5): 1175-1186.

122. GUO Y, XU Q, DANIELE C, et al. CRISPR inversion of CTCF sites alters genome topology and enhancer/promoter function. Cell, 2015, 162(4): 900-910.

123. SHEN H, XU W, GUO R, et al. Suppression of enhancer over-activation by a rack7-histone demethylase complex. Cell, 2016, 165(2): 331-342.

124. WANG Q C, ZHENG Q, TAN H, et al. TMCO1 is an ER Ca^{2+} load-activated Ca^{2+}(CLAC) channel. Cell, 2016, 165: 1454-1466.

125. LI X, CUI X L, WANG J Q, et al. Generation and application of mouse-rat allodiploid embryonic stem cells. Cell, 2016, 164(1-2): 279–292.

126. CHEN Y, YU J, NIU Y, et al. Modeling rett syndrome using talen-edited MECP2

mutant cynomolgus monkeys. Cell, 2017, 169(5): 945-955.
127. KE Y, XU Y, CHEN X, et al. 3D chromatin structures of mature gametes and structural reprogramming during mammalian embryogenesis. Cell, 2017, 170(2): 367-381.
128. HUANG F, CHEN Y G. Regulation of TGF-beta receptor activity. Cell Biosci, 2012, 2: 9.
129. WU M, WANG L, LI Q, et al. The MTA family proteins as novel histone H3 binding proteins. Cell Biosci, 2013, 3(1): 1.
130. LIAO X, DENG W, LU Y, et al. Sp1 plays an important role in regulating the transcription of ZNF313. Cell Biol Int, 2010, 34(9): 901-905.
131. ZHU H, QIN Y, ZHU C Q. Neuronal cell surface ATP synthase mediates synthesis of extracellular ATP. Cell Biol Int Rep, 2011, 35: 81-86.
132. JIANG Z, FAN Q, ZHANG Z, et al. SENP1 deficiency promotes ER stress-induced apoptosis by increasing XBP1 SUMOylation. Cell Cycle, 2012, 116: 1118-1122.
133. WU J, HE Z, WANG D L, et al. Depletion of JMJD5 sensitizes tumor cells to microtubule-destabilizing agents by altering microtubule stability. Cell Cycle, 2016, 15(21): 2980-2991.
134. LI Z, ZHANG W, CHEN Y, et al. Impaired DNA double-strand break repair contributes to th eage-associated rise of genomic instability in humans. Cell Death Differ, 2016, 23(11): 1765-1777.
135. YE Y, LI M, GU L, et al. Chromatin remodeling during in vivo neural stem cells differentiating to neurons in early Drosophila embryos. Cell Death Differ, 2017, 24(3): 409-420.
136. DU Y, LIU Z, CAO X, et al. Nucleosome eviction along with H3K9ac deposition enhances Sox2 binding during human neuroectodermal commitment. Cell Death Differ, 2017, 24(6): 1121-1131.
137. HUANG Y, CHEN J, LU C, et al. HDAC1 and Klf4 interplay critically regulates human myeloid leukemia cell proliferation. Cell Death Dis, 2014, 5e1491: 1-12.
138. LI X, GAO D, WANG H, et al. Negative feedback loop between p66Shc and ZEB1 regulates fibrotic EMT response in lung cancer cells. Cell Death Dis, 2015, 6: e1708.
139. HUANG L, ZHU P, XIA P, et al. WASH has a critical role in NK cell cytotoxicity through Lck-mediated phosphorylation. Cell Death Dis, 2016, 7: e2301.
140. ZHAI L L, WU R M, HAN W H, et al. miR-127 enhances myogenic cell differentiation by targeting S1PR3. Cell Death Dis, 2017, 8(3): e2707.
141. QIAO Y B, WANG X J, WANG R, et al. AF9 promotes hESC neural differentiation through recruiting TET2 to neurodevelopmental gene loci for methylcytosine

hydroxylation. Cell Discov, 2015, 1: 15017.
142. LIU W, LIU X, WANG C, et al. Identification of key factors conquering developmental arrest of somatic cell cloned embryos by combining embryo biopsy and single-cell sequencing. Cell Discov, 2016, 7(2): 16010.
143. JIA Y H, LI P S, FANG L, et al. Negative regulation of DNMT3A de novo DNA methylation by frequently overexpressed UHRF family proteins as a mechanism for widespread DNA hypomethylation in cancer. Cell Discov, 2016, 2: 16007.
144. HE S, CHEN J, ZHANG Y, et al. Sequential EMT-MET induces neuronal conversion through Sox2. Cell Discov, 2017, 3: 17017.
145. JIN C Y, LI J, CHRISTOPHER D G, et al. Histone demethylase UTX-1 regulates C. elegans life span by targeting the Insulin/IGF-1 signaling pathway. Cell Metab, 2011, 14: 161-172.
146. HOU L, TANG D, CHEN D, et al. A Systems approach to reverse engineer lifespan extension by dietary restriction. Cell Metab, 2016, 23(3): 529-540.
147. LIU Y N, LI N M, LI T, et al. POU homeodomain protein OCT1 modulates islet 1 expression during cardiac differentiation of P19CL6 cells. Cell Mol Life Sci, 2011, 68(11): 1969-1982.
148. LIU Z H, YANG G, ZHAO T C, et al. Small ncRNA expression and regulation under hypoxia in neural progenitor cells. Cell Mol Neurobiol, 2011, 31(1): 1-5.
149. CHEN Y, HU W, LU Y, et al. A TALEN-based specific transcript knock-down of PIWIL2 suppresses cell growth in HepG2 tumor cell. Cell Prolif, 2014, 47(5): 448-456.
150. LI P, GAO S, WANG L, et al. ABH2 couples regulation of rDNA Transcription with DNA alkylation Repair. Cell Rep, 2013, 4(4): 817-829.
151. ZHANG W, LIU Y, SUN N, et al. Integrating genomic, epigenomic, and transcriptomic features reveals modular signatures underlying poor prognosis in ovarian cancer. Cell Rep, 2013, 4(3): 542-553.
152. SU M, HAN D L, JEROME D B, et al. Evolution of alu towards enhancers. Cell Rep, 2014, 7(2): 376-435.
153. NIE J W, JIANG M Y, ZHANG X T, et al. Post-transcriptional regulation of Nkx2-5 by RHAU in heart development. Cell Rep, 2015, 13: 723-732.
154. CHEN J, CHEN X, LI M, et al. Hierarchical Oct4 binding in concert with primed epigenetic rearrangements during somatic cell reprogramming. Cell Rep, 2016, 14: 1540-1554.
155. SUN J, LUAN Y, XIANG D, et al. The 11S proteasome subunit PSME3 is a positive feedforward regulator of NF-κB and important for host defense against bacterial

pathogens. Cell Rep, 2016, 14: 737-749.
156. MA C, KARWACKI-NEISIUS V, TANG H, et al. Nono, a bivalent domain factor, regulates erk signaling and mouse embryonic stem cell pluripotency. Cell Rep, 2016, 17(4): 997-1007.
157. CHRISTOPHER D G, HUANG Y, DOU X Y, et al. Impact of dietary interventions on noncoding rna networks and mrnas encoding chromatin-related factors. Cell Rep, 2017, 18(12): 2957-2968.
158. ZHU L L, ZHAO T, HUANG X, et al. Gene expression profiles and metabolic changes in embryonic progenitor cells under low oxygen. Cell Rep, 2011, 13: 113-120.
159. TANG H, SHA H, SUN H, et al. Tracking induced pluripotent stem cells-derived neural stem cells in the central nervous system of rats and monkeys cellular reprogramming. Cell Rep, 2013, 15: 435-442.
160. HUANG J, CHEN T, LIU X, et al. More synergetic cooperation of Yamanaka factors in induced pluripotent stem cells than in embryonic stem cells. Cell Res, 2009, 19(10): 1127-1138.
161. DONG L H, WEN J K, MIAO S B, et al. Baicalin inhibits PDGF-BB-stimulated vascular smooth muscle cell proliferation through suppressing PDGFRbeta-ERK signaling and increase in p27 accumulation and prevents injury-induced neointimal hyperplasia. Cell Res, 2010, 20: 1252-1262.
162. YANG Z, JIANG J, STEWART M D, et al. AOF1 is a histone H3K4 demethylase possessing demethylase-independent repression activity. Cell Res, 2010, 20(3): 276-287.
163. QIU J, SHI G, JIA Y, et al. The X-linked mental retardation gene PHF8 is a histone demethylase involved in neuron differentiation. Cell Res, 2010, 20(8): 908-918.
164. LIN H, WANG Y, XU Y, et al. Coordinated regulation of active and repressive histone methylations by a dual specificity histone demethylase ceKDM7A from C. elegans. Cell Res, 2010, 20: 899-907.
165. YANG Y, HU L, CHEN C D, et al. Structural insights into a dual specificity histone demethylase ceKDM7A from C. elegans. Cell Res, 2010, 20: 886-898.
166. ZHU Z, WANG Y, CHEN P A, et al. PHF8 is a histone H3K9me2 demethylase regulating rRNA synthesis. Cell Res, 2010, 20: 794-801.
167. HUANG C, XIANG Y, JING N, et al. Dual specificity histone demethylase KIAA1718 (KDM7A) regulates neural differentiation through FGF4. Cell Res, 2010, 20: 154-165.
168. CHEN L Y, WANG D K, WU Z T, et al. Molecular basis of the first cell fate

determination in mouse embryogenesis. Cell Res, 2010, 20: 982-993.

169. YU L, WANG Y, HUANG S, et al. Structural insights into a novel histone demethylase PHF8. Cell Res, 2010, 20(2): 166-173.

170. ZHENG B, HAN M, SHU Y N, et al. HDAC2 phosphorylation-dependent Klf5 deacetylation and RARα acetylation induced by RAR agonist switch the transcription regulatory programs of p21 in VSMCs. Cell Res, 2011, 21: 1487-1508.

171. ZHANG J, GAO Q, LI P, et al. S phase-dependent interaction with DNMT1 dictates the role of UHRF1 but not UHRF2 in DNA methylation maintenance. Cell Res, 2011, 21(12): 1723-1739.

172. WANG H L, MU Y J, CHEN D H. Effective gene silencing in Drosophila ovarian germline by artificial microRNAs. Cell Res, 2011, 21: 700-703.

173. LI W, ZHAO X Y, WAN H F, et al. iPS cells generated without c-Myc have active Dlk1-Dio3 region and are capable of producing full-term mice through tetraploid complementation. Cell Res, 2011, 21(3): 550-553.

174. TONG M, LV Z, LIU L, et al. Mice generated from tetraploid complementation competent iPS cells show similar developmental features as those from ES cells but are prone to tumorigenesis. Cell Res, 2011, 21(11): 1634-1637.

175. SHENG C, ZHENG Q, WU J, et al. Generation of dopaminergic neurons directly from mouse fibroblasts and fibroblast-derived neural progenitors. Cell Res, 2012, 22(4): 769-772.

176. SHENG C, ZHENG Q, WU J, et al. Direct reprogramming of Sertoli cells into multipotent neural stem cells by defined factors. Cell Res, 2012, 22(1): 208-218.

177. QIAO Y B, ZHU Y, SHENG N Y, et al. AP2g regulates neural and epidermal fates determination as a BMP downstream target. Cell Res, 2012, 22: 1546-1561.

178. GUO X, LIU Q, WANG G, et al. microRNA-29b is a novel mediator of Sox2 function in the regulation of somatic cell reprogramming. Cell Res, 2013, 23: 142-156.

179. ZHANG Q, QI S, XU M, et al. Structure-function analysis reveals a novel mechanism for regulation of histone demethylase LSD2/AOF1/KDM1b. Cell Res, 2013, 23(2): 225-241.

180. LIU Y, QIAO N, ZHU S S, et al. A novel bayesian network inference algorithm for integrative analysis of heterogeneous deep sequencing data. Cell Res, 2013, 23(3): 440-443.

181. ZHAO Y, LI X, CAI M, et al. XBP-1u suppresses autophagy by promoting the degradation of FoxO1 in cancer cells. Cell Res, 2013, 23(4): 491-507.

182. ZHANG Q, QI S K, XU M C, et al. Structure-function analysis reveals a novel

mechanism for regulation of histone demethylase LSD2/AOF1/KDM1b. Cell Res, 2013, 23: 225-241.

183. WAN H F, HE Z Q, DONG M Z, et al. Parthenogenetic haploid embryonic stem cells produce fertile mice. Cell Res, 2013, 23(11): 1330.

184. ZHAO X, YANG Y, SUN B F, et al. FTO-dependent demethylation of N6-methyladenosine regulates mRNA splicing and is required for adipogenesis. Cell Res, 2014, 24(12): 1403-1419.

185. GOU L T, DAI P, YANG J H, et al. Pachytene piRNAs instruct massive mRNA elimination during late spermiogenesis. Cell Res, 2014, 24(6): 680-700.

186. HU S, ZHU W, ZHANG L F, et al. MicroRNA-155 broadly orchestrates inflammation-induced changes of microRNA expression in breast cancer. Cell Res, 2014, 24(2): 254-257.

187. ZHU T, ROUNDTREE I A, WANG P, et al. Crystal structure of the YTH domain of YTHDF2 reveals mechanism for recognition of N6-methyladenosine. Cell Res, 2014, 24(12): 1493-1496.

188. REN R, LIU H, WANG W, et al. Structure and domain organization of drosophila tudor. Cell Res, 2014, 24: 1146-1149.

189. WU Y, ZHOU H, FAN X, et al. Correction of a genetic disease by crispr-cas9-mediated gene editing in mouse spermatogonial stem cells. Cell Res, 2015, 25: 67-79.

190. SHU J, ZHANG K, ZHANG M, et al. GATA family members as inducers for cellular reprogramming to pluripotency. Cell Res, 2015, 25(2): 169-180.

191. LI C Y, KAN L J, CHEN Y, et al. Ci antagonizes Hippo signaling in the somatic cells of the ovary to drive germline stem cell differentiation. Cell Res, 2015, 25: 1152-1170.

192. CHEN W Y, QIAN W, WU G, et al. Three-dimensional human facial morphologies as robust aging markers. Cell Res, 2015, 25(5): 574-587.

193. ZHANG P, KANG J Y, GOU L T, et al. MIWI and piRNA-mediated cleavage of messenger RNAs in mouse testes. Cell Res, 2015, 25(2): 193-207.

194. ZHANG Y, GU L F, HOU Y F, et al. Integrative genome-wide analysis reveals HLP1, a novel RNA-binding protein, regulates plant flowering by targeting alternative polyadenylation. Cell Res, 2015, 25: 864-876.

195. WAN H, FENG C, TENG F, et al. One-step generation of p53 gene biallelic mutant cynomolgus monkey via the CRISPR/Cas system. Cell Res, 2015, 25(2): 258-261.

196. ZHONG C, ZHANG M, YIN Q, et al. Generation of human haploid embryonic stem cells from parthenogenetic embryos obtained by microsurgical removal of male pronucleus. Cell Res, 2016, 26: 743-746.

197. ZHONG C, XIE Z, YIN Q, et al. Parthenogenetic haploid embryonic stem cells efficiently support mouse generation by oocyte injection. Cell Res, 2016, 26: 131-134.
198. CHEN K, ZHANG J, GUO Z Q, et al. Loss of 5-hydroxymethylcytosine is linked to gene body hypermethylation in kidney cancer. Cell Res, 2016, 26: 103-118.
199. YE J, GE J, ZHANG X, et al. Pluripotent stem cells induced from mouse neural stem cells and small intestinal epithelial cells by small molecule compounds. Cell Res, 2016, 26(1): 34-45.
200. ZHOU C L, YANG X Q, SUN Y Y, et al. Comprehensive profiling reveals mechanisms of SOX2-mediated cell fate specification in human ESCs and NPCs. Cell Res, 2016, 26: 171-189.
201. LI Z, WAN H, FENG G, et al. Birth of fertile bimaternal offspring following intracytoplasmic injection of parthenogenetic haploid embryonic stem cells. Cell Res, 2016, 26(1): 135-138.
202. SHI X L, GAO Y, YAN Y, et al. Improved survival of porcine acute liver failure by a bioartificial liver device implanted with induced human functional hepatocytes. Cell Res, 2016, 26(2): 206-216.
203. WANG Z W, HU B Q, SUN Q Y, et al. Oocyte-expressed yes-associated protein is a key activator of the early zygotic genome in mouse. Cell Res, 2016, 26: 275-287.
204. ZHAO D, GUAN H, ZHAO S, et al. YEATS2 is a selective histone crotonylation reader. Cell Res, 2016, 26(5): 629-632.
205. YAN H, ZHU S, SONG C, et al. Bone morphogenetic protein (BMP) signaling regulates mitotic checkpoint protein levels in human breast cancer cells. Cell Signal, 2012, 24: 961-968.
206. LI Z, FEI T, ZHANG J, et al. BMP4 signaling acts via dual specificity phosphatase 9 to control erk activity in mouse embryonic stem cells. Cell Stem Cell, 2012, 10: 171-182.
207. ZHANG R R, CUI Q Y, KIYOHITO M, et al. Tet1 regulates adult hippocampal neurogenesis and cognition. Cell Stem Cell, 2013, 13: 237-245.
208. WU Y, LIANG D, WANG Y, et al. Correction of a genetic disease in mouse via use of CRISPR-Cas9. Cell Stem Cell, 2013, 13: 659-662.
209. FANG R, LIU K, ZHAO Y, et al. Generation of naive induced pluripotent stem cells from rhesus monkey fibroblasts. Cell Stem Cell, 2014, 15(4): 488-496.
210. GUO F, LI X, LIANG D, et al. Active and passive demethylation of male and female pronuclear DNA in the mammalian zygote. Cell Stem Cell, 2014, 15: 447-458.
211. LIU H, CHEN Y, NIU Y, et al. TALEN-mediated gene mutagenesis in rhesus and cynomolgus monkeys. Cell Stem Cell, 2014, 14(3): 323-328.

212. LI X, ZUO X, JING J, et al. Small-molecule-driven direct reprogramming of mouse fibroblasts into functional Neurons. Cell Stem Cell, 2015, 17(2): 195-203.
213. ZHONG C, YIN Q, XIE Z, et al. CRISPR-cas9-mediated genetic screening in mice with haploid embryonic stem cells carrying a guide RNA library. Cell Stem Cell, 2015, 17: 221-232.
214. CHEN T, HAO Y J, ZHANG Y, et al. m6A RNA methylation is regulated by microRNAs and promotes reprogramming to pluripotency. Cell Stem Cell, 2015, 16(3): 289-301.
215. WANG Y, HE L, DU Y, et al. The long noncoding RNA lncTCF7 primes self-renewal of liver cancer stem cells through activation of Wnt signaling. Cell Stem Cell, 2015, 16(4): 366-375.
216. ZHU Z, LI C, ZENG Y, et al. PHB Associates with the HIRA complex to control an epigenetic-metabolic circuit in human ESCs. Cell Stem Cell, 2017, 20(2): 274-289.
217. LIU X, DONG L, ZHANG X, et al. Identification of p100 protein target promoters by chromatin immunoprecipitation guided ligation and selection (ChIP-GLAS). Cell Mol Immunology, 2011, 8: 88-91.
218. ZHANG L F, JIANG S, LIU M F. MicroRNA regulation and analytical methods in cancer cell metabolism. Cell Mol Life Sci, 2017 (5): 1-13.
219. QIAO Y B, YANG X F, JING N H. Epigenetic regulation of early neural fate commitment. Cell Mol Life Sci, 2016, 73(7): 1399-1411.
220. YANG P, LU Y, JIANG X, et al. Potential role of RING finger protein 166 (RNF166), a member of an ubiquitin ligase subfamily, involved in regulation of T cell activation. Cent Eur J Immunol, 2013, 38(1): 15-22.
221. QIN L, NIU Q, LI X, et al. CTCF mediates long-range interaction between silencer Sis and enhancer Ei and inhibits VJ rearrangement in pre-B cells. Cent Eur J Immunol, 2013, 38(3): 349-354.
222. CHI L, FAN B, FENG D, et al. The dorsoventral patterning of human forebrain follows an activation/transformation model. Cereb Cortex, 2017, 27(5): 2941-2954.
223. SHAO N, ZHANG K, CHEN Y, et al. Preparation and characterization of DNA aptamer based spin column for enrichment and separation of histones. Chem Commun, 2012, 48: 6684-6686.
224. TANG H L, SUN H P, WU X, et al. Detection of neural stem cells function in rats with traumatic brain injury by manganese-enhanced magnetic resonance imaging. Chin Med J (Engl), 2011, 124: 1848-1853.
225. XIE L Q, SUN H P, WANG T, et al. Reprogramming of adult human neural stem cells into induced pluripotent stem cells. Chin Med J (Engl), 2013, 126: 1138-1143.

226. LIU Z, SUN Y. Role of tip60 tumor suppressor in DNA repair pathway. Chinese Sci Bull, 2011, 56 (12): 1212-1215.
227. ZHAO Q Y, LEI P J, ZHANG X, et al. Global histone modification profiling reveals the epigenomic dynamics during malignant transformation in a four-stage breast cancer model. Clin Epigenetics, 2016, 8: 34.
228. WANG H, ZHANG P, LIU L, et al. Hierarchical organization and regulation of the hematopoietic stem cell osteoblastic niche. Crit Rev Oncol Hematol, 2013, 85: 1-8.
229. YANG N, XU R M. Structure and function of the BAH domain in chromatin biology. Crit Rev Biochem Mol Biol, 2013, 48: 211-221.
230. XIA L X, ZHENG X D, ZHENG W J, et al. The niche-dependent feedback loop enerates a BMP activity gradient to determine the germline stem cell fate. Curr Biology, 2012, 22: 515-521.
231. YANG Z, SUN N, WANG S, et al. Feud or friend? The role of the mir-17-92 cluster in tumorigenesis. Curr Genomics, 2010, 11: 129-135.
232. HOU L, HUANG J L, CHRISTOPHER D G, et al. Systems biology in aging: linking the old and the young. Curr Genomics, 2012, 13(7): 558-565.
233. LIN Q, CUI P, DING F, et al. Replication-associated mutational pressure (RMP) governs strand-biased compositional asymmetry (SCA) and Gene Organization in animal mitochondrial genomes. Curr Genomics, 2011, 12: 28-36.
234. LI G H, DANNY R. Chromatin higher-order structures and gene regulation. Curr Opin Genet Dev, 2010, 21: 175-186.
235. ZHU P, LI G H. Structural insights of nucleosome and the 30-nm chromatin fiber. Curr Opin Genet Dev, 2016, 36: 106-115.
236. YAO Y, YANG Y, ZHU W G. Sirtuins: nodes connecting aging, metabolism and tumorigenesis. Curr Pharm Design, 2014, 20(11): 1614-1624.
237. ZHAO Y L, FANG X L, WANG Y, et al. Comprehensive Analysis for histone acetylation of human colon cancer cells treated with a novel HDAC inhibitor. Curr Pharm Design, 2014, 20: 1866-1873.
238. XUE Y, GAO X, CAO J, et al. A summary of computational resources for protein phosphorylation. Curr Protein Pept Sc, 2010, 11: 485-496.
239. Ren J, Gao X, Liu Z, et al. Computational analysis of phosphoproteomics: progresses and perspectives. Curr Protein Pept Sc, 2011, 12: 591-601.
240. CHEN M, LI T, LIN S, et al. Association of Interleukin 6 gene polymorphisms with genetic susceptibilities to spastic tetraplegia in males: a case-control study. Cytokine, 2013, 61(3): 826-830.
241. LIU H J, WU Z T, SHI X L, et al. Atypical PKC, regulated by Rho GTPases and

Mek/Erk, phosphorylates Ezrin during eight-cell embryo compaction. Dev Biol, 2013, 375: 13-22.

242. HU B, ZHANG W, FENG X, et al. Zebrafish eaf1 suppresses foxo3b expression to modulate transcriptional activity of gata1 and spi1 in primitive hematopoiesis. Dev Biol, 2014, 388: 81-93.

243. MA P, XIA Y, MA L, et al. Xenopus Nkx6. 1 and Nkx6. 2 are required for mid-hindbrain boundary development. Dev Genes Evol, 2013, 223: 253-259.

244. FANG H, YANG Y, LI C L, et al. Transcriptome analysis of early organogenesis in human embryos. Dev Cell, 2010, 19: 174-184.

245. ZHAO S, GOU L T, ZHANG M, et al. piRNA-triggered MIWI ubiquitination and removal by APC/C in late spermatogenesis. Dev Cell, 2013, 24(1): 13-25.

246. NING G Z, LIU X L, DAI M M, et al. MicroRNA-92a upholds Bmp signaling by targeting noggin3 during pharyngeal cartilage formation. Dev Cell, 2013, 24(3): 283-295.

247. YU Z L, ZHOU X, WANG W J, et al. Dynamic phosphorylation of CENP-A at Ser68 orchestrates its cell cycle-dependent deposition at centromeres. Dev Cell, 2015, 32: 68-81.

248. PENG G D, SUO S B, CHEN J, et al. Spatial transcriptome for the molecular annotation of lineage fates and cell identity in mid-gastrula mouse embryo. Dev Cell, 2016, 36（6）: 681-697.

249. ZHANG K J, LI L Y, HUANG C Y, et al. Distinct functions of BMP4 during different stages of mouse ES cell neural commitment. Development, 2010, 137: 2095-2105.

250. LIU B, LIU Y F, DU Y R, et al. Cbx4 regulates the proliferation of thymic epithelial cells and thymus function. Development, 2013, 140: 780-788.

251. LI L Y, LIU C, STEFFEN B, et al. Location of transient ectodermal progenitor potential in mouse development. Development, 2013, 140: 4533-4543.

252. LIU J X, ZHANG D W, XIE X W, et al. Eaf1 and Eaf2 negatively regulate canonical wnt/beta-catenin signaling. Development, 2013, 140: 1067-1078.

253. LUO W, ZHAO X, JIN H W, et al. Akt1 signaling coordinates BMP signaling and β-catenin activity to regulate second heart field progenitor development. Development, 2015, 142: 732-742.

254. LIU C, SONG Z, WANG L, et al. Sirt1 regulates acrosome biogenesis by modulating autophagic flux during spermiogenesis in mice. Development, 2016, 144(3): 441-451.

255. SHA Q Q, DAI X X, DANG Y, et al. A MAPK cascade couples maternal mRNA translation and degradation to meiotic cell cycle progression in mouse oocytes.

Development, 2017, 144(3): 452-463.
256. XIAO Q, ZHANG G X, WANG H J, et al. A p53 based genetic tracing system to follow postnatal cardiomyocyte expansion in heart regeneration. Development, 2017, 144: 580-589.
257. HUANG C Y, CHEN J, ZHANG T, et al. The dual histone demethylase KDM7A promotes neural induction in early chick embryos. Developmental Dynamics, 2010, 239(12): 3350-3357.
258. HE X Y, WANG X B, ZHANG R, et al. Investigation of mycoplasma pneumoniae infection in pediatric population from 12025 cases with respiratory infection. Diagn Microbiol Infect Dis, 2013, 75: 22-27.
259. WU C, LI A, LENG Y, et al. Histone deacetylase inhibition by sodium valproate regulates polarization of macrophage subsets. DNA Cell Biol, 2012, 31(4): 592-599.
260. WANG H Y, ZHOU W, ZHENG Z X, et al. The HDAC inhibitor depsipeptide transactivates the p53/p21 pathway by inducing DNA damage. DNA Repair (Amst), 2012, 11(2): 146-156.
261. ZHANG Y, LI C, LI H, et al. miR-378 activates the pyruvate-PEP futile cycle and enhances lipolysis to ameliorate obesity in mice. E Bio Med, 2016, 5: 93-104.
262. ZHU Q Q, SONG L, PENG G D, et al. The transcription factor Pou3f1 promotes neural fate commitment via activation of neural lineage genes and inhibition of external signaling pathways. eLife, 2014, 3: e02224.
263. ZHAO J, LIN Q, KEVIN J K, et al. Ngn1 inhibits astrogliogenesis through induction of miR-9 during neuronal fate specification. eLife, 2015, 4: 1-11.
264. FANG Q L, CHEN P, WANG M Z, et al. Human cytomegalovirus IE1 protein alters the higher-order chromatin structure by targeting the acidic patch of the nucleosome. eLife, 2016, 5: e11911.
265. JIANG S, ZHANG L F, ZHANG H W, et al. A novel miR-155/miR-143 cascade controls glycolysis by regulating hexokinase 2 in breast cancer cells. EMBO J, 2012, 31(8): 1985-1998.
266. YAO X, TANG Z Y, YIN J W, et al. The mediator subunit MED23 couples H2B mono-ubiquitination to transcriptional control and cell fate determination. EMBO J, 2015, 34(23): 2885-2902.
267. ZHANG L F, LOU J T, LU M H, et al. Suppression of miR-199a maturation by HuR is crucial for hypoxia-induced glycolytic switch in hepatocellular carcinoma. EMBO J, 2015, 34(21): 2671-2685.
268. SUN S G, ZHENG B, HAN M, et al. miR-146a and Krüppel-like factor 4 form a feedback loop to participate in vascular smooth muscle cell proliferation. EMBO Rep,

2011, 12: 56-62.
269. CHEN K S, LONG Q, WANG T, et al. Gadd45a is a heterochramatin relaxer that enhances iPS cell generation. EMBO Rep, 2016, 17: 1641-1656.
270. FENG G H, TONG M, XIA B L, et al. Ubiquitously expressed genes participate in cell-specific functions via alternative promoter usage. EMBO Rep, 2016, 17: 1304-1313.
271. YANG Y, ZHOU C, WANG Y, et al. The E3 ubiquitin ligase RNF114 and TAB1 degradation are required for maternal-tozygotic transition. EMBO Rep, 2017, 18(2): 205-216.
272. WANG N L, ZHANG P, GUO X J, et al. A protein differentially expressed in immature rat ovarian development, is required for normal primordial follicle assembly and development. Endocrinology, 2011, 152(3): 1024-1035.
273. ZHANG Z, YU J. Does the genetic code have a eukaryotic origin? Enomics Proteomics Bioinformatics, 2013, 11: 41-55.
274. JEROME D B, CHRISTOPHER D G, WU G, et al. Epigenomics and the regulation of aging. Epigenomics, 2013, 5(2): 205-227.
275. MA X T, WANG Y W, MICHAEL Q Z, et al. DNA methylation data analysis and its application to cancer research. Epigenomics, 2013, 5(3): 301-316.
276. CHEN J, WANG G, WANG L, et al. Curcumin p38-dependently enhances the anticancer activity of valproic acid in human leukemia cells. Eur J Pharm Sci, 2010, 41: 210-218.
277. WANG D P, SU Y, LEI H X, et al. Transposon-derived and satellite-derived repetitive sequences play distinct roles in the intron size expansion among mammalian genomes. Evolutionary Bioinformatics, 2012, 8: 301-319.
278. XIA Y, FANG H, ZHANG J, et al. Endoplasmic reticulum stress-mediated apoptosis in imatinib-resistant leukemic K562-r cells triggered by AMN107 combined with arsenic trioxide. Exp Biol Med (Maywood), 2013, 238: 932-942.
279. LU W, ZHANG Y, LIU D, et al. Telomeres-structure, function, and regulation. Exp Cell Res, 2013, 319(2): 133-141.
280. LI Z, CHEN Y G. Functions of BMP Signaling in Embryonic Stem Cell Fate determination. Exp Cell Res, 2013, 319(2): 113-119.
281. SHEN H, XU W, LAN F. Histone lysine demethylases in mammalian embryonic development. Exp Mol Med, 2017, 49(4): e325.
282. DONG F, HAO S, MA S, et al. A novel lymphoid progenitor cell population (LSKlow) is restricted by p18INK4c. Exp Hematol, 2016, 44: 874-885.
283. XIA X, HEN W Y, JOSEPH M D, et al. Molecular and phenotypic biomarkers of

284. ZHENG Z, LI L, LIU X, et al. 5-Aza-2'-deoxycytidine reactivates gene expression via degradation of pRb. FASEB J, 2012, 26(1): 449-459.

285. WANG F, XIONG L, HUANG X, et al. DNA demethylation regulates the expression of miR-210 in neural progenitor cells subjected to hypoxia. FASEB J, 2012, 279: 4318-4326.

286. WANG H B, LIU G H, ZHANG H, et al. Sp1 and c-Myc regulate transcription of BMI-1 in nasopharyngeal carcinoma. FASEB J, 2013, 280: 2929-2944.

287. ZHENG Z, YANG J, ZHAO D, et al. Downregulated adaptor protein p66(Shc) mitigates autophagy process by low nutrient and enhances apoptotic resistance in human lung adenocarcinoma A549 cells. FASEB J, 2013, 280(18): 4522-4530.

288. SHEN C, WANG D, LIU X, et al. SET7/9 regulates cancer cell proliferation by influencing beta-catenin stability. FASEB J, 2015, 29(30): 4313-4323.

289. GAO X, GE L, SHAO J, et al. Tudor-SN interacts with and co-localizes with G3BP in stress granules under stress conditions. FEBS Lett, 2010, 584: 3525-3532.

290. LI G H, ZHU P. Structure and organization of chromatin fiber in the nucleus. FEBS Lett, 2015, 589: 2893-2904.

291. ZHANG L, LU D Y, MA W Y, et al. Age-related changes in the localization of DNA methyltransferases during meiotic maturation in mouse oocytes. Fertil Steril, 2011, 95(4): 1531-1534.

292. LIU F, ZHANG H X, WU G, et al. Sequence variation and expression analysis of seed dormancy- and germination-associated ABA- and GA-related genes in rice cultivars. Front Plant Sci, 2011, 2: 17.

293. GUO T T, FANG Y. Functional organization and dynamics of the cell nucleus. Front Plant Sci, 2014, 5: 378.

294. GE Y, LIU J, ZENG M, et al. Identification of WOX family genes in Selaginella kraussiana for studies on stem cells and regeneration in lycophytes. Front Plant Sci, 2016, 7: 93.

295. CAO J, XIAO Z, CHEN B, et al. Differential effects of recombinant fusion proteins TAT-OCT4 and TAT-NANOG on adult human fibroblasts. Front Biol, 2010, 5: 424-430.

296. SHI L L, FANG Y. Histone variants: making structurally and functionally divergent nucleosomes and linkers in chromatin. Front Biol, 2011, 6: 93-101.

297. XIE Z H, SHENG N Y, JING N H. BMP signaling pathway and spinal cord development. Front Biol, 2012, 7: 24-29.

298. ZHANG S J, LUO Z H, SHI G, et al. TPP1 as a versatile player at the ends of

chromosomes. Front Biol, 2014, 9(3): 225-233.
299. LIU W F, ZHAO Y H, CUI P, et al. Thousands of novel transcripts identified in mouse cerebrum, testis, and ES cells based on ribo-minus RNA sequencing. Front Genet, 2011, 2: 93.
300. FANG Y Y, ZHAO J H, LIU S W, et al. CMV2b-AGO interaction is required for the suppression of rdr-dependent antiviral silencing in arabidopsis. Front Microbiol, 2016, 7: 1329.
301. ZHANG K, ZHU L L, FAN M. Oxygen, a key factor regulating cell behavior during neurogenesis and cerebral diseases. Mol Neurosci, 2011, 4: 1-11.
302. LIU S, SUN X, WANG M, et al. A microRNA 221- and 222-mediated feedback loop maintains constitutive activation of NF κ B and STAT3 in colorectal cancer cells. Gastroenterology, 2014, 147: 847-859.
303. LI L Y, JING N H. Pluripotent stem cell studies elucidate the underlying mechanisms of early embryonic development. Genes, 2011, 2: 298-312.
304. WANG W C, QIN Z Y, FENG Z X, et al. Identifying differentially spliced genes from two groups of RNA-Seq samples. Genes, 2013, 518(1): 164-170.
305. LIU H, WANG J Y, HUANG Y, et al. Structural basis for methylarginine. dependent recognition of aubergine by tudor. Genes & Dev, 2010, 24: 1876-1881.
306. HU H, LIU Y, WANG M, et al. Structure of a CENP, a Histone H4 heterodimer in complex with chaperone HJURP. Genes & Dev, 2011, 25: 901-906.
307. CHIEN Y, SCUOPPO C, WANG X W, et al. Control of the senescence-associated secretory phenotype by NF-kappa B promotes senescence and enhances chemosensitivity. Genes & Dev, 2011, 25(20): 2125-2136.
308. QIU Y, LIU L, ZHAO C, et al. Combinatorial readout of unmodified H3R2 and acetylated H3K14 by the tandem PHD finger of MOZ reveals a regulatory mechanism for HOXA9 transcription. Genes & Dev, 2012, 26: 1376-1391.
309. OZLEM A, AGUSTIN C, ZENG T Y, et al. The atypical E2F family member E2F7 couples the p53 and Rb pathways during cellular senescence. Genes & Dev, 2012, 26(14): 1546-1557.
310. RATNAKUMAR K, DUARTE L F, LEROY G, et al. ATRX-mediated chromatin association of histone variant macroH2A1 regulates α-globin expression. Genes & Dev, 2012, 26(5): 433-438.
311. SU H C, WANG C L, WANG M, et al. Structural basis for allosteric stimulation of Sir2 activity by Sir4 binding. Genes & Dev, 2013, 27: 64-73.
312. CHEN P, ZHAO J C, WANG Y, et al. H3. 3 actively marks enhancers and primes gene transcription via opening higher-ordered chromatin. Genes & Dev, 2013, 27:

2109-2124.

313. FANG J N, WEI Y, LIU Y T, et al. Structural transitions of centromeric chromatin regulate the cell cycle-dependent recruitment of ENP-N. Genes & Dev, 2015, 29: 1058-1073.

314. YAN R R, HE L, LEI Z W, et al. SCF(JFK) is a bona fide E3 ligase for ING4 and a potent promoter of the angiogenesis and metastasis of breast cancer. Genes & Dev, 2015, 29: 672-685.

315. MOHAMED N D, LIANG Z Y, WANG Q, et al. 3CPET: finding co-factor complexes from ChIA-PET data using a hierarchical Dirichlet process. Genome Biol, 2015, 16: 288.

316. ZHANG X L, WU J, WANG J, et al. Integrative epigenomic analysis reveals unique epigenetic signatures involved in unipotency of mouse female germline stem cells. Genome Biol, 2016, 17: 162.

317. HUANG Y, YU X M, SUN N, et al. Single-cell level spatial gene expression in the embryonic neural differentiation niche. Genome Res, 2015, 25(4): 570-581.

318. LI T, WANG S, WU R, et al. Identification of long non-protein coding RNAs in chicken skeletal muscle using next generation sequencing. Genomics, 2012, 99(5): 292-298.

319. LI R J, GUO W L, GU J, et al. Chromatin state and microRNA determine different gene expression dynamics responsive to TNF stimulation. Genomics, 2012, 100(5): 297-302.

320. LIU P Y, DOU X Y, PENG G D, et al. Genome-wide analysis of histone acetylation dynamics during mouse embryonic stem cell neural differentiation. Genomics Data, 2015, 5: 15-16.

321. SONG L, SUN N, PENG G D, et al. Genome-wide ChIP-seq and RNA-Seq analyses of Pou3f1 during mouse pluripotent stem cell neural fate commitment. Genomics Data, 2015, 5: 375-377.

322. HAO W, HU Z Q, ZHANG Z, et al. Strand-biased gene distribution in bacteria is related to both horizontal gene transfer and strand-biased nucleotide composition. Genomics Proteomics, 2012, 10(4): 186-196.

323. CUI P, LIN Q, DING F, et al. Transcript-centric mutations in human genomes. GPB, 2012, 10(1): 11-22.

324. MA L N, NIE L H, ZHANG B, et al. A RNA-Seq-based gene expression profiling of radiation-induced tumorigenic mammary epithelial cells. GPB, 2012, 10: 326-335.

325. CUI P, LIU W F, ZHAO Y H, et al. The association between H3K4me3 and antisense transcription. GPB, 2012, 10(2): 74-81.

326. ZHANG Z, YU J. The pendulum model for genome compositional dynamics: from the four nucleotides to the twenty amino acids. GPB, 2012, 10 (4): 175-180.
327. ZHANG M, YANG C, LIU H, et al. Induced pluripotent stem cells are sensitive to DNA damage. GPB, 2013, 11(5): 320-326.
328. WANG C Q, ZHANG M Q, ZHANG Z H. Computational identification of active enhancers in model organisms. GPB, 2013, 11(3): 142-150.
329. WU J Y, XIAO J F, ZHANG Z, et al. Ribogenomics: the science and knowledge of RNA. GPB, 2014, 12: 57-63.
330. RASHID F, SHAH A, SHAN G. Long Non-coding RNAs in the Cytoplasm. GPB, 2016, 14(2): 73-80.
331. CHEN H Z, LIU Q F, LI L, et al. The orphan receptor TR3 suppresses intestinal tumorigenesis in mice by downregulating Wnt signalling. Gut, 2011, 61: 714-724.
332. QIAO A, LIANG J, KE Y, et al. Mouse patatin-like phospholipase domain-containing 3 influences systemic lipid and glucose homeostasis. Hepatology, 2011, 54: 509-521.
333. LU C L, LIN L, TAN H P, et al. Fragile X premutation RNA is sufficient to cause primary ovarian insufficiency in mice. Hum Mol Genet, 2012, 21: 5039-5047.
334. ZHANG W X, CHENG Y, LI Y J, et al. A feed-forward mechanism involving Drosophila fragile X mental retardation protein triggers a replication stress-induced DNA damage response. Hum Mol Genet, 2014, 23: 5188-5196.
335. LAI D P, TAN S, KANG Y N, et al. Genome-wide profiling of polyadenylation sites reveals a link between selective polyadenylation and cancer metastasis. Hum Mol Genet, 2015, 24(12): 3410-3417.
336. ADWAIT S, ZHANG Y, AIAOTU M, et al. SCT promoter methylation is a highly discriminative biomarker for lung and many other cancers. IEEE Life Sci Lett, 2015, 1(3): 30-33.
337. WANG S, XIA P, CHEN Y, et al. Natural killer-like B cells prime innate lymphocytes against microbial infection. Immunity, 2016, 45(1): 131-144.
338. YANG P, LU Y, LI M, et al. Identification of RNF114 as a novel positive regulatory protein for T cell activation. Immunobiology, 2014, 219(6): 432-439.
339. WANG X, HUANG Y, ZHAO J, et al. Suppression of PRMT6-mediated arginine methylation of p16 protein potentiates its ability to arrest A549 cell proliferation. Int J Biochem Cell Biol, 2012, 44(12): 2333-2341.
340. GU B, ZHU W G. Surf the post-translational modification network of p53 regulation. Int J Biol Sci, 2012, 8(5): 672-684.
341. LIU H, LONG J, ZHANG P, et al. Elevated b-arrestin1 expression correlated with

risk stratification in acute lymphoblastic leukemia. Int J Hematol, 2011, 93: 494-501.
342. LI Z, ZHU W G. Targeting Histone Deacetylases for Cancer Therapy: From Molecular Mechanisms to Clinical Implications. Int J Biol Sci, 2014, 10(7): 757-770.
343. CHI H, RONG R, LEI C, et al. Effects of thoracic epidural blockade on mortality of patients with idiopathic dilated cardiomyopathy and heart failure. Int J Biol Sci, 2011, 150(3): 350-351.
344. ZHU P, LI G H. Higher-Order Structure of the 30-nm Chromatin Fiber Revealed by Cryo-EM. Iubmb Life, 2016, 68: 873-878.
345. GU C, GONG H, ZHANG Z, et al. Association of interleukin-10 gene promoter polymorphisms with recurrent pregnancy loss: a meta-analysis. J Assist Reprod Genet, 2016, 33(7): 907-917.
346. LI H X, HAN M, BERNIER M, et al. Kruppel-like factor 4 promotes differentiation by transforming growth factor-beta receptor-mediated Smad and p38 MAPK signaling in vascular smooth muscle cells. J Biol Chem, 2010, 285: 17846-17856.
347. LU X, SHI Y, LU Q, et al. Requirement for lamin B receptor and its regulation by importin {beta} and phosphorylation in nuclear envelope assembly during mitotic exit. J Biol Chem, 2010, 285(43): 33281-33293.
348. SUN H Q, LI D, CHEN S, et al. Zili inhibits transforming growth factor-beta signaling by interacting with Smad4. J Biol Chem, 2010, 285(6): 4243-4250.
349. DONG L, ZHANG X, FU X, et al. PTB-associated splicing factor (PSF) functions as a repressor of STAT6-mediated Ig epsilon gene transcription by recruitment of HDAC1. J Biol Chem, 2011, 286: 3451-3459.
350. ZHOU B O, ZHOU J Q. Recent transcription-induced histone H3K4 methylation inhibits gene reactivation. J Biol Chem, 2011, 286(40): 34770-34776.
351. LI J, ZHOU F, ZHAN D, et al. A novel histone H4 arginine 3 methylation-sensitive histone H4 binding activity and transcriptional regulatory function for signal recognition particle subunits SRP68 and SRP72. J Biol Chem, 2012, 287(48): 40641-40651.
352. WU Z T, YANG M, LIU H J, et al. Role of nuclear receptor coactivator 3 (ncoa3) in pluripotency maintenance. J Biol Chem, 2012, 287: 38295-38304.
353. SHI J H, ZHENG B, CHEN S, et al. Retinoic acid receptor alpha mediates all-trans retinoic acid-induced Klf4 gene expression by regulating Klf4 promoter activity in vascular smooth muscle cells. J Biol Chem, 2012, 287: 10799-10811.
354. LI J, CHU M, WANG S, et al. Identification and characterization of nardilysin as a novel dimethyl H3K4 binding protein involved in transcriptional regulation. J Biol Chem, 2012, 287(13): 10089-10098.

355. GAO X, ZHAO X, ZHU Y, et al. Tudor staphylococcal nuclease (Tudor-SN) participates in small ribonucleoprotein (snRNP) assembly via interacting with symmetrically dimethylated sm proteins. J Biol Chem, 2012, 287: 18130-18141.

356. ZHANG T, ZHU Q Q, XIE Z H, et al. The zinc-finger transcription factor Ovol2 acts downstream of the BMP pathway to regulate the cell fate decision between neuroectoderm and mesendoderm. J Biol Chem, 2013, 288: 6166-6177.

357. SONG C, ZHU S, WU C, et al. HDAC10 suppresses cervical cancer metastasis through inhibition of matrix metalloproteinase (MMP) 2 and 9 expression. J Biol Chem, 2013, 288(39): 28021-28033.

358. CHEN L, WEI T, SI X, et al. Lysine acetyltransferase GCN5 potentiates the growth of non-small cell lung cancer via promotion of E2F1, Cyclin D1 and Cyclin E1 Expression. J Biol Chem, 2013, 288(20): 14510-14521.

359. SHI F T, KIM H, LU W, et al. Ten-eleven translocation 1 (Tet1) is regulated by O-linked N-acetylglucosamine transferase (Ogt) for target gene repression in mouse embryonic stem cells. J Biol Chem, 2013, 288(29): 20776-20784.

360. WAN M, LIANG J, XIONG Y, et al. The trithorax group protein Ash2l is essential for pluripotency and maintaining open chromatin in embryonic stem cells. J Biol Chem, 2013, 288(7): 5039-5048.

361. TANG M, LI Y, ZHANG X, et al. Structural maintenance of chromosomes flexible hinge domain containing 1 (SMCHD1) promotes non-homologous end joining and inhibits homologous recombination repair upon DNA damage. J Biol Chem, 2014, 289(49): 34024-34032.

362. LIU Y, KIM H, LIANG J, et al. The death-inducer obliterator 1 (Dido1) gene regulates embryonic stem cell self-renewal. J Biol Chem, 2014, 289(8): 4778-4786.

363. FENG C, LIU Y, WANG G Q, et al. Crystal structures of the human rna demethylase Alkbh5 reveal basis for substrate recognition. J Biol Chem, 2014, 289(17): 11571-11583.

364. QIAO Y B, WANG R, YANG X F, et al. Dual roles of histone h3 lysine 9 acetylation in human embryonic stem cell pluripotency and neural differentiation. J Biol Chem, 2015, 290(4): 2508-2520.

365. TAN F Z, QIAN C, TANG K, et al. Inhibition of tgf-β signaling can substitute for oct4 in reprogramming and maintain pluripotency. J Biol Chem, 2015, 290(7): 4500-4511.

366. SU C, ZHANG C, ADIAM T, et al. Tudor Staphylococcal Nuclease (Tudor-SN), a Novel Regulator Facilitating G1/S Phase Transition, Acting As a Co-activator of E2F-1 in Cell Cycle Regulation. J Biol Chem, 2015, 290: 7208-7220.

367. SHI S S, LIU K, CHEN Y H, et al. Competitive inhibition of lysine acetyltransferase 2B by a small motif of the adenoviral oncoprotein E1A. J Biol Chem, 2016, 291(27): 14363-14372.
368. LIN Q, GENG J, MA K. RASSF1A, APC, ESR1, ABCB1 and HOXC9, but not p16INK4A, DAPK1, PTEN and MT1G genes were frequently methylated in the stage I non-small cell lung cancer in China. J Cancer Res Clin Oncol, 2009, 135: 1675-1684.
369. GAO X, PAN W S, DAI H, et al. CARM1 activates myogenin gene via PCAF in the early differentiation of TPA-induced rhabdomyosarcoma-derived cells. J Cell Biochem, 2010, 110: 162-170.
370. DAI J P, LU J Y, ZHANG Y, et al. Jmjd3 activates Mash1 gene in RA-induced neuronal differentiation of P19 cells. J Cell Biochem, 2010, 110: 1457-1463.
371. ZHU S, LI Y, HUANG B, et al. TSA-induced JMJD2B downregulation is associated with Cyclin B1-dependent survivin degradation and apoptosis in LNCap cells. J Cell Biochem, 2012, 113(7): 2375-2382.
372. LI B H, JIA Z Q, WANG T, et al. Interaction of Wnt/β-catenin and Notch signaling in the early stage of cardiac differentiation of P19CL6 cells. J Cell Biochem, 2012, 113(2): 629-639.
373. FONG K W, LI Y, WANG W, et al. Whole-genome screening identifies proteins localized to distinct nuclear bodies. J Cell Biol, 2013, 203(1): 149-164.
374. HU X Q, LI T, ZHANG C G, et al. GATA4 regulates ANF expression synergistically with Sp1 in a cardiac hypertrophy model. J Cell Mol Med, 2011, 15(9): 1865-1877.
375. ZHAO J C, ZHANG L X, ZHANG Y, et al. The differential regulation of Gap43 gene in the neuronal differentiation of P19 cells. J Cell Physiology, 2012, 227: 1645-2653.
376. WANG X K, ZHU S, LI Y, et al. A H3k4 Methyltransferase, regulates the tnfalpha-stimulated activation of genes downstream of Nf-Kappab. J Cell Sci, 2012, 125: 4058-4066.
377. FONG K W, LEUNG J W, LI Y, et al. MTR120/KIAA1383, a novel microtubule-associated protein, promotes microtubule stability and ensures cytokinesis. J Cell Sci, 2013, 126(Pt 3): 825-837.
378. FENG X, LUO Z, JIANG S, et al. The telomere-associated homeobox-containing protein TAH1/HMBOX1 participates in telomere maintenance in ALT cells. J Cell Sci, 2013, 126(Pt 17): 3982-3989.
379. CAI W Y, WEI T Z, LUO Q C, et al. Wnt/β-catenin pathway represses let-7 microRNAs expression via transactivation of Lin28 to augment breast cancer stem

cell expansion. J Cell Sci, 2013, 126: 2877-2889.
380. LI Y, FONG K W, TANG M, et al. Fam118B, a newly identified component of Cajal bodies, is required for Cajal body formation, snRNP biogenesis and cell viability. J Cell Sci, 2014, 127(Pt 9): 2029-2039.
381. TANG M, LI Y, ZHANG Y, et al. Disease mutant analysis identifies a new function of DAXX in telomerase regulation and telomere maintenance. J Cell Sci, 2015, 128(2): 331-341.
382. WANG Z, HUANG M, MA X, et al. REV1 promotes PCNA monoubiquitylation through interacting with ubiquitylated RAD18. J Cell Sci, 2016, 129(6): 1223-1233.
383. YAN J, ZHANG H, LIU Y, et al. Germline deletion of huntingtin causes male infertility and arrested spermiogenesis in mice. J Cell Sci, 2016, 129(3): 492-501.
384. ZHU P, WANG Y, HE L, et al. ZIC2-dependent OCT4 activation drives self-renewal of human liver cancer stem cells. J Clin Invest, 2015, 125(10): 3795-3808.
385. WANG Q, MA S, SONG N, et al. Stabilization of histone demethylase PHF8 by USP7 promotes breast carcinogenesis. J Clin Invest, 2016, 126(6): 2205-2220.
386. XIA P, WANG S, DU Y, et al. Insulin-InsR signaling drives multipotent progenitor differentiation toward lymphoid lineages. J Exp Med, 2015, 212(13): 2305-2321.
387. GAO Y, WANG X, HAN J, et al. The novel OCT4 spliced variant OCT4B1 can generate three protein isoforms by alternative splicing into OCT4B. J Genet Genomics, 2010, 37: 461-465.
388. BAI M, LIANG D, WANG Y, et al. Spermatogenic cell-specific gene mutation in mice via CRISPR-Cas9. J Genet Genomics, 2016, 43: 289-296.
389. ZHU K, WANG X, JU L G, et al. Wdr82 negatively regulates cellular antiviral response by mediating Traf3 polyubiquitination in multiple cell lines. J Immunol, 2015, 195: 5358-5366.
390. LI J, CHAI Q Y, ZHANG Y, et al. Mycobacterium tuberculosis Mce3E suppresses host innate immune responses by targeting ERK1/2 signaling. J Immunol, 2015, 194: 3756-3767.
391. ZHANG Y, GAO X, ZHI L, et al. The synergetic antibacterial activity of Ag islands on ZnO (Ag/ZnO) heterostructure nanoparticles and its mode of action. J Inorg Biochem, 2014, 130: 74-83.
392. BAI M, WU Y, LI J S. Generation and application of mammalian haploid embryonic stem cells. J Internal Med, 2016, 280: 236-245.
393. LIU Z, JIANG Y, HOU Y, et al. The IκB family member Bcl-3 stabilizes c-Myc in colorectal cancer. J Mol Cell Biol, 2013, 5: 280-282.
394. LI L Y, SONG L, LIU C, et al. Ectodermal progenitors derived from epiblast stem

cells by inhibition of nodal signaling. J Mol Cell Biol, 2015, 7(5): 455-465.
395. LIU Z, CAO W, XU L, et al. The histone H3 lysine-27 demethylase Jmjd3 plays a critical role in specific regulation of Th17 cell differentiation. J Mol Cell Biol, 2015, 7: 505-516.
396. LIU Y, LIU S, YUAN S, et al. Chromodomain protein CDYL is required for transmission/restoration of repressive histone marks. J Mol Cell Biol, 2017, 9(3): 1-17.
397. BI D, CHEN M, ZHANG X, et al. The association between sex-related interleukin-6 gene polymorphisms and the risk for cerebral palsy. J Neuroinflammation, 2014, 11(1): 1-12.
398. CUI Z Y, HOU J J, CHEN X L, et al. The Profile of mitochondrial proteins and their phosphorylation signaling network in INS-1 b cells. J Proteome Res, 2010, 9(6): 2898-2908.
399. ZENG M, LIANG S, ZHAO S, et al. Identifying mRNAs bound by human RBMY protein in the testis. J Reprod Dev, 2011, 57(1): 107-112.
400. LIU X K, ZHANG X R, ZHONG Q, et al. Low expression of PTK6/Brk predicts poor prognosis in patients with laryngeal squamous cell carcinoma. J Transl Med, 2013, 11: 59.
401. ZOU Q F, DU J K, ZHANG H, et al. Anti-tumour activity of longikaurin A (LK-A), a novel natural diterpenoid, in nasopharyngeal carcinoma. J Transl Med, 2013, 11: 200.
402. ZHANG M F, LI X R, DENG Z Q, et al. Structural biology of the arterivirus nsp11 endoribonucleases. J Virol, 2017, 91: e01309-16.
403. LIU L, LUO G Z, YANG W, et al. Activation of the imprinted Dlk1-Dio3 region correlates with pluripotency levels of mouse stem cells. J Biol Chem, 2010, 285(25): 19483-19490.
404. QIAO Q, LI Y, CHEN Z, et al. The structure of NSD1 reveals an autoregulatory mechanism underlying histone H3K36 methylation. J Biol Chem, 2011, 286: 8361-8368.
405. LIU Z T, LIN X W, CAI Z P, et al. Global identification of SMAD2 target genes reveals a role for multiple co-regulatory factors in zebrafish early gastrulas. J Biol Chem, 2011, 286(32): 28520-28532.
406. ZHANG Y, YANG X, GUI B, et al. Corepressor protein CDYL functions as a molecular bridge between polycomb repressor complex 2 and repressive chromatin mark Trimethylated Histone Lysine 27. J Biol Chem, 2011, 286: 42414-42425.
407. WAN X Y, HU B, LIU J X, et al. Zebrafish mll gene is essential for hematopoiesis. J Biol Chem, 2011, 286(38): 33345-33357.

408. CAI R, YU T, HUANG C, et al. SUMO-specific protease 1 regulates mitochondrial biogenesis through PGC-1a. J Biol Chem, 2012, 287(53): 44464-44470.
409. YU Z P, FAN D S, GUI B, et al. Neurodegeneration-associated TDP-43 interacts with fragile X mental retardation protein (FMRP)/Staufen (STAU1) and regulates SIRT1 expression in neuronal cells. J Biol Chem, 2012, 287: 22560-22572.
410. ZHANG X S, ZHANG B L, GAO J D, et al. Regulation of the microRNA 200b (miRNA-200b) by transcriptional regulators PEA3 and ELK-1 protein affects expression of Pin1 protein to control anoikis. J Biol Chem, 2013, 288(45): 32742-32752.
411. ZHANG Q, LIU X, GAO W, et al. Differential regulation of the ten-eleven translocation (TET) family of dioxygenases by O-linked β-N-acetylglucosamine transferase (OGT). J Biol Chem, 2014, 289(9): 5986-5996.
412. SHI G, WU M, FANG L, et al. PHD finger protein 2 (PHF2) represses ribosomal RNA gene transcription by antagonizing PHD finger protein 8 (PHF8) and recruiting methyltransferase SUV39H1. J Biol Chem, 2014, 289(43): 29691-29700.
413. QI Y, ZUO Y, EDWARD T H, et al. An essential role of small ubiquitin-like modifier (SUMO)-specific Protease 2 in Myostatin expression and myogenesis. J Biol Chem, 2014, 289(6): 3288-3293.
414. LIU B, WANG T, MEI W, et al. Small ubiqiutin-like modifier(SUMO) protein-specific protease 1 De-SUMOylates sharp-1 protein and controls adipocyte differentiation. J Biol Chem, 2014, 289(32): 22358-22364.
415. QI S K, WANG Z Q, LI P S, et al. Non-germline restoration of genomic imprinting for a subset of imprinted genes in ubiquitin-like PHD and RING finger domain-containing 1 (Uhrf1) null mouse embryonic stem cells. J Biol Chem, 2015, 290(22): 14181-14191.
416. SUN G N, HU Z R, MIN Z Y, et al. Small C-terminal domain phosphatase 3 dephosphorylates the linker sites of receptor-regulated Smads (R-Smads) to ensure transforming growth factor beta (TGFbeta)-mediated germ layer induction in xenopus embryos. J Biol Chem, 2015, 290(28): 17239-17249.
417. XIAO L H, MEI W, CHENG J. SUMOylation attenuates human β-Arrestin 2 Inhibition of IL-1R/TRAF6 signaling. J Biol Chem, 2015, 290(4): 1927-1935.
418. LIU X, CHEN Z, OUYANG G, et al. ELL protein-associated factor 2 (EAF2) inhibits transforming growth factor beta signaling through a direct interaction with Smad3. J Biol Chem, 2015, 290(43): 25933-25945.
419. MEI Z C, ZHANG D W, HU B, et al. FBX032 targets c-Myc for proteasomal degradation and inhibits c-Myc activity. J Biol Chem, 2015, 290(26): 16202-16214.

420. SONG L, CHEN J, PENG G D, et al. Dynamic heterogeneity of brachyury in mouse epiblast stem cells mediates distinct respond to extrinsic bmp signaling. J Biol Chem, 2016, 291(29): 15212-15225.

421. DING G, CHEN P, ZHANG H, et al. Regulation of ubiquitin-like with plant homeodomain and ring finger domain 1 (UHRF1) protein stability by heat shock protein 90 chaperone machinery. J Biol Chem, 2016, 291(38): 20125-20135.

422. WANG L, CHEN W L, XU X, et al. Activin/Smad2-induced H3K27me3 reduction is crucial to initiate mesendoderm differentiation of human embryonic stem cells. J Biol Chem, 2016, 292(4): 1339-1350.

423. HUANG H D, LI Y J, KEITH E S, et al. AGO3 Slicer activity regulates mitochondria–nuage localization of Armitage and piRNA amplification. J Cell Biol, 2014, 206: 217-230.

424. CHEN L, TONG J, XIAO L, et al. YUCCA-mediated auxin biogenesis is required for cell fate transition occurring during de novo root organogenesis in Arabidopsis. J Exp Bot, 2016, 67: 4011-4013.

425. ZHU X H, ZHANG H, QIAN M X, et al. The significance of low PU. 1 expression in patients with acute promyelocyticleukemia. J Hematol Oncol, 2012, 5: 22.

426. QIAN M X, JIN W, ZHU X H, et al. Structurally differentiated cis-elements that interact with PU. 1 are functionally distinguishable in acute promyelocytic leukemia. J Hematol Oncol, 2013, 6: 25.

427. CAI T X, SHU Q B, LIU P, et al. Characterization and relative quantification of phospholipids based on methylation and stable isotopic labeling. J Lipid Res, 2016, 57(3): 388-397.

428. HAN G, LI J, LIU Y, et al. The hydroxylation activity of Jmjd6 is required for its homo-oligomerization, journal of cellular biochemistry. J Cell Biochem, 2012, 113: 1663-1670.

429. JI S, ZHANG L, HUI L. Cell fate conversion: Direct induction of hepatocyte-like cells from fibroblasts. J Cell Biochem, 2013, 114(2): 256-265.

430. SUN M, LIAO B, TAO Y, et al. Calcineurin-NFAT signaling controls somatic cell reprogramming in a stage-dependent manner. J Cell Physiol, 2015, 231(5): 1151-1162.

431. LU C C, ZHANG K, ZHANG Y, et al. Preparation and characterization of vorinostat-coated beads for profiling of novel target proteins. J Chromatogr A, 2014, 1372: 34-41.

432. WANG Q, MA S, SONG N, et al. Stabilization of histone demethylase PHF8 by USP7 promotes breast carcinogenesis. J Clin Invest, 2016, 126: 2205-2220.

433. LIU X L, LIN D Y, MA W Y. Quantitative analysis of intracellular calcium and

mitochondrial kinetic fluorescence changes in GSNO-induced thymocyte early apoptosis. J Fluoresc, 2011, 21(3): 1285-1292.

434. LI Y, LIU X Y, HUANG L, et al. Potential coexistence of both bacterial and eukaryotic small RNA biogenesis and functional related protein homologs in Archaea. J Genet Genomics, 2010, 37: 493-503.

435. LI Y H, KANG X J, WANG Q. HSP70 decreases receptor-dependent phosphorylation of Smad2 and blocks TGF-beta-induced epithelial-mesenchymal transition. J Genet Genomics, 2011, 38(3): 111-116.

436. CHEN P, ZHAO J C, LI G H. Histone variants in development and diseases. J Genet Genomics, 2013, 40: 355-365.

437. LIU X L, MA Y Q, ZHANG C W, et al. Nodal Promotes mir206 expression to control convergence and extension movements during zebrafish gastrulation. J Genet Genomics, 2013, 40: 515-521.

438. LI Z, WANG L, WANG Y, et al. Generation of an LncRNA Gtl2-GFP reporter for rapid assessment of pluripotency in mouse induced pluripotent stem cells. J Genet Genomics, 2015, 42(3): 125-128.

439. HUANG H Y, WU Q. CRISPR double cutting through the labyrinthine architecture of 3D genomes. J Genet Genomics, 2016, 43: 273-288.

440. LU Y, ZHOU X Y, JIN Z G, et al. Resolving the genetic heterogeneity of prelingual hearing loss within one family: Performance comparison and application of two targeted next generation sequencing approaches. J Hum Genet, 2014, 59(11): 599-607.

441. YING Z Z, MEI M, ZHANG P Z, et al. Histone arginine methylation by PRMT7 controls germinal center formation via regulating Bcl6 transcription. J Immunol, 2015, 195(4): 1538-1547.

442. WANG W, GAO L, WANG X, et al. Modulation of the poliovirus receptor expression in malignant lymphocytes by epigenetic alterations. J Immunol, 2011, 34: 353-361.

443. LIU X C, WANG P F, FU J H, et al. Two-photon fluorescence real-time imaging on the development of early mouse embryo by stages. J Microsc, 2011, 241(2): 212-218.

444. CAI R, GU J, SUN H, et al. Induction of SENP1 in myocardium contributes to abnormities of mitochondria and cardiomyopathy. J Mol Cell Cardiol, 2015, 79: 115-122.

445. ZHANG Y, YANG Z, YANG Y, et al. Production of transgenic mice by random recombination of targeted genes in female germline stem cells. J Mol Cell Cardiol, 2011, 3: 132-141.

446. XU C H, LV X W, ERIC Z C, et al. Genome-wide roles of Foxa2 in directing liver

specification. J Mol Cell Cardiol, 2012, 4(6): 420-422.

447. LI X, WANG J Q, WANG L Y, et al. Co-participation of paternal and maternal genomes before the blastocyst stage is not required for full-term development of mouse embryos. J Mol Cell Cardiol, 2015, 7(5): 486-488.

448. SHUAI L, WANG Y, DONG M, et al. Durable pluripotency and haploidy in epiblast stem cells derived from haploid embryonic stem cells in vitro. J Mol Cell Cardiol, 2015, 7(4): 326-337.

449. WANG Q, HE J, MENG L, et al. A proteomics analysis of rat liver membrane skeletons: the investigation of actin- and cytokeratin-based protein components. J Proteome Res, 2010, 9(1): 22-29.

450. ZHANG Y A, MA X T, ADWAIT S, et al. Validation of SCT methylation as a hallmark biomarker for lung cancers. J Thorac Oncol, 2016, 11(3): 346-360.

451. LONG J, LIU S, LI K, et al. High proportion of CD34+/CD38-cells is positively correlated with poor prognosis in newly diagnosed childhood acute lymphoblastic leukemia. Leuk Lymphoma, 2014, 55: 611-617.

452. ZHENG Y, ZHANG H, WANG Y, et al. Loss of Dnmt3b accelerates MLL-AF9 leukemia progression. Leukemia, 2016, 30: 1-12.

453. LIU T H, TANG Y J, HUANG Y, et al. Expression of the fetal hematopoiesis regulator FEV indicates leukemias of prenatal origin. Leukemia, 2016, 31(5): 1079-1086.

454. GAO X N, LIN J, GAO L, et al. MicroRNA-193b regulates c-Kit proto-oncogene and represses cell proliferation in acute myeloid leukemia. Leuk Res, 2011, 35: 1226-1232.

455. LIU X K, LI Q, XU L H, et al. Expression and clinical significance of SIAH in laryngeal squamous cell carcinoma. Med Oncol, 2013, 30: 485.

456. INKMAN A B, SIMMER F, MA K. Whole-genome DNA methylation profiling using MethylCap-Seq. Methods, 2010, 52: 232-236.

457. JIN Y, GUO H S. Transgene-induced gene silencing in plants. Methods Mol Biol, 2015, 1287: 105-117.

458. ZHONG C, LI J S. Efficient generation of gene-modified mice by haploid embryonic stem cell-mediated semi-cloned technology. Methods Mol Biol, 2017, 1498: 121-133.

459. CHEN L Y, ZHANG Y, ZHANG Q, et al. Mitochondrial localization of telomeric protein TIN2 links telomere regulation to metabolic control. Mol Cell, 2012, 47(6): 839-850.

460. ZHOU T, XIONG J, WANG M, et al. Structural basis for hydroxymethylcytosine recognition by the SRA domain of UHRF2. Mol Cell, 2014, 54: 879-886.

461. FANG L, ZHANG L, WEI W, et al. A methylation-phosphorylation switch determines Sox2 stability and function in esc maintenance or differentiation. Mol Cell, 2014, 55(4): 537-551.

462. YANG Y, YIN X, YANG H. Histone demethylase LSD2 Acts as an E3 ubiquitin ligase and inhibits cancer cell growth through promoting proteasomal degradation of OGT. Mol Cell, 2015, 58: 47-59.

463. LI Y, SABARI B R, PANCHENKO T, et al. Molecular coupling of histone crotonylation and active transcription by AF9 YEATS domain. Mol Cell, 2016, 62(2): 181-193.

464. WANG Y N, ZHANG N, ZHANG L Y, et al. Autophagy regulates chromatin ubiquitination in DNA damage response through elimination of SQSTM1/p62. Mol Cell, 2016, 63(1): 34-48.

465. LI W, CHEN P, YU J, et al. FACT remodels the tetranucleosomal unit of chromatin fibers for gene transcription. Mol Cell, 2016, 64: 120-133.

466. ZHENG H, HUANG B, ZHANG B J, et al. Resetting epigenetic memory by reprogramming of histone modifications in mammals. Mol Cell, 2016, 63(6): 1066-1079.

467. ZHANG W H, XIA W K, WANG Q J, et al. Isoform switch of TET1 regulates dna demethylation and mouse development. Mol Cell, 2016, 64(6): 1062-1073.

468. XIONG J, ZHANG Z, CHEN J, et al. Cooperative action between SALL4A and TET Proteins in stepwise oxidation of 5-methylcytosine. Mol Cell, 2016, 64(5): 913-925.

469. XIAO W, ADHIKARI S, DAHAL U, et al. Nuclear m6A reader YTHDC1 regulates mRNA splicing. Mol Cell, 2016, 61(4): 507-519.

470. ZHANG L, WANG G, WANG L, et al. VPA inhibits breast cancer cell migration by specifically targeting HDAC2 and down-regulating Survivin. Mol Cell Biochem, 2012, 361: 39-45.

471. SHI H J, WEN J K, MIAO S B, et al. KLF5 and hhLIM cooperatively promote proliferation of vascular smooth muscle cells. Mol Cell Biochem, 2012, 367: 185-194.

472. CHEN J, LUO Q, YUAN Y, et al. Pygo2 associates with MLL2 histone methyltransferase (HMT) and GCN5 histone acetyltransferase (HAT) complexes to augment Wnt target gene expression and breast cancer stem-like cell expansion. Mol Cell Biol, 2010, 30: 5621-5635.

473. ZHOU B O, WANG S S, XU L X, et al. SWR1 complex poises heterochromatin boundaries for antisilencing activity propagation. Mol Cell Biol, 2010, 30(10): 2391-2400.

474. WANG S S, ZHOU B O, ZHOU J Q. Histone H3 lysine 4 hepermethylation prevents aberrant nucleosome remodeling at the PHO5 promoter. Mol Cell Biol, 2011, 31(15): 3171-3181.
475. TONG X J, LI Q J, DUAN Y M, et al. Est1 protects telomeres and inhibits subtelomeric Y-element recombination. Mol Cell Biol, 2011, 31(6): 1263-1274.
476. LIU X L, NING G Z, MENG A M, et al. MicroRNA-206 regulates cell movements during zebrafish gastrulation by targeting prickle1a and Regulating c-Jun N-terminal kinase 2 phosphorylation. Mol Cell Biol, 2012, 32: 2934-2942.
477. TAO R, XUE H, ZHANG J, et al. Deacetylase rpd3 facilitates checkpoint adaptation by preventing rad53 overactivation. Mol Cell Biol, 2013, 33(21): 4212-4224.
478. JIANG S, ZHAO L, LU Y, et al. Piwil2 inhibits Keratin 8 degradation through promoting p38-induced phosphorylation to resist Fas-mediated apoptosis. Mol Cell Biol, 2014, 34(21): 3928-3938.
479. ZHAO X, LU S S, NIE J W, et al. Phosphoinositide-dependent kinase 1 and mTORC2 synergistically maintain postnatal heart growth and heart function in mice. Mol Cell Biol, 2014, 34: 1966-1975.
480. CHEN Z, LIU X, MEI Z C, et al. Eaf2 suppresses hypoxia-induced factor 1a transcriptional activity by disrupting its interaction with coactivator CBP/p300. Mol Cell Biol, 2014, 34(6): 1085-1099.
481. ZHANG W, JI W, LIU X, et al. ELL inhibits E2F1 transcriptional activity by enhancing E2F1 deacetylation via recruiment of histone deacetylase 1. Mol Cell Biol, 2014, 34(4): 765-775.
482. CHEN S, WANG C, SUN L, et al. RAD6 promotes homologous recombination repair by activating the autophagy-mediated degradation of heterochromatin protein HP. Mol Cell Biol, 2015, 35(2): 406-416.
483. AN H, YANG L, WANG C, et al. Interactome analysis reveals a novel role for RAD6 in the regulation of proteasome activity and localization in response to DNA damage. Mol Cell Biol, 2017, 37(6): e00419- e00416.
484. HONG S J, HUANG Y, CAO Y Q, et al. Approaches to uncovering cancer diagnostic and prognostic molecular signatures. Mol Cell Oncol, 2014, 1(2): e957981.
485. REN J, JIANG C, GAO X, et al. PhosSNP for systematic analysis of genetic polymorphisms that influence protein phosphorylation. Mol Cell Proteomics, 2010, 9: 623-634.
486. SONG C, YE M, LIU Z, et al. Systematic analysis of protein phosphorylation networks from phosphoproteomic data. Mol Cell Proteomics, 2012, 11(10): 133-141.
487. GAN H, CAI T, LIN X, et al. Integrative proteomic and transcriptomic analyses

reveal multiple post-transcriptional regulatory mechanisms of mouse spermatogenesis. Mol Cell Proteomics, 2013, 12(5): 1144-1157.
488. WANG L, WANG G, YANG D, et al. Euphol arrests breast cancer cells at the G1 phase through the modulation of cyclin D1, p21 and p27 expression. Mol Med Rep, 2013, 8: 1279-1285.
489. LIN J, CHEN G, GU L, et al. Phylogenetic affinity of tree shrews to Glires is attributed to fast evolution rate. Mol Phylogenet. Evol, 2014, 71: 193-200.
490. LIU Y, LIU Q, YAN Q Q, et al. Nucleolus-tethering System (NoTS) reveals the assembly of photobodies follows a self-organization model. Mol Biol Cell, 2014, 25: 1366-1373.
491. LIU Z, MA Q, CAO J, et al. GPS-PUP: computational prediction of pupylation sites in prokaryotic proteins. Mol Bio Systems, 2011, 7: 2737-2740.
492. LIU Z, MA Q, CAO J, et al. GPS-YNO2: Computational prediction of tyrosine nitration sites in proteins. Mol Bio Systems, 2011, 7: 1197-1204.
493. XIE P, LIU Y, LI Y, et al. MIROR: a method for cell-type specific microRNA occupancy rate prediction. Mol Bio systems, 2014, 10(6): 1377-1384.
494. GU J, CHEN Y, HUANG H, et al. Gene module based regulator inference identifying miR-139 as a tumor suppressor in colorectal cancer. Mol Bio Systems, 2014, 10(12): 3249-3254.
495. GUO B H, FENG Y, ZHANG R, et al. Bmi-1 promotes invasion and metastasis, and its elevated expression is correlated with an advanced stage of breast cancer. Molecular Cancer, 2011, 10(1): 10.
496. WANG J, ZHANG W, JI W, et al. The Von Hippel-Lindau protein suppresses androgen receptor activity. Mol Endocrinology, 2014, 28(2): 239-248.
497. LI Z, LI B, DONG A. The arabidopsis transcription factor AtTCP15 regulates endoreduplication by modulating expression of key cell-cycle genes. Mol Plant, 2012, 5: 270-280.
498. YANG C, GU L, DENG D. Distinct susceptibility of induction of methylation of p16(ink4a) and p19(arf) CpG islands by X-radiation and chemical carcinogen in mice. Mutat Res, 2014, 768: 42-50.
499. LI D, ZHANG W, ZHENG S, et al. Surveillance study of candidemia in cancer patients in North China. Mydical Mycology, 2013, 51(4): 378-384.
500. XIANG H, ZHU J, CHEN Q, et al. Single base-resolution methylome of the silkworm reveals a sparse epigenomic map. Nat Biotechnol, 2010, 28: 516-552.
501. LI W, TENG F, LI T, et al. Simultaneous generation and germline transmission of multiple gene mutations in rat using CRISPR-Cas systems. Nat Biotechnol, 2013,

31(8): 684-686.

502. ZHAO Y, YANG J, LIAO W, et al. Cytosolic FoxO1 is essential for the induction of autophagy and suppressor activity. Nat Cell Biol, 2010, 12(7): 665-675.

503. LIU J, HAN Q K, PENG T R, et al. The oncogene c-Jun impedes somatic cell reprogramming. Nat Cell Biol, 2015, 7(7): 856-867.

504. XIONG X, PANCHENKO T, YANG S, et al. Selective recognition of histone crotonylation by double PHD fingers of MOZ and DPF2. Nat Chem Biol, 2016, 12(12): 1111-1118.

505. LIU X, GAO Q, LI P, et al. UHRF1 targets DNMT1 for DNA methylation through cooperative binding of hemi-methylated DNA and methylated H3K9. Nat Commun, 2013, 4: 1563.

506. CHEN W Z, LIU Y, ZHU S S, et al. Improved nucleosome-positioning algorithm iNPS for accurate nucleosome positioning fromsequencing data. Nat Commun, 2014, 18(5): 4909.

507. XUE Y, ZHENG X D, HUANG L, et al. Organizer-derived Bmp2 is required for the formation of a correct Bmp activity gradient during embryonic development. Nat Commun, 2014, 5: 3766.

508. WU R, LI H, ZHAI L, et al. MicroRNA-431 accelerates muscle regeneration and ameliorates muscular dystrophy by targeting Pax7 in mice. Nat Commun, 2015, 6: 7713.

509. ZHU P, WANG Y, DU Y, et al. C8orf4 negatively regulates self-renewal of liver cancer stem cells via suppression of NOTCH2 signaling. Nat Commun, 2015, 6: 7122.

510. CHENG J, YANG H, FANG J, et al. Molecular mechanism for USP7-mediated DNMT1 stabilization by acetylation. Nat Commun, 2015, 6: 7023.

511. SUN J, WEI H, XU J, et al. Histone H-mediated epigenetic regulation controls germline stem cell self-renewal by modulating H4K16 acetylation. Nat Commun, 2015, 6: 8856.

512. XIA P Y, WANG S, XIONG B Q, et al. IRTKS negatively regulates antiviral immunity through PCBP2 sumoylation-mediated MAVS degradation. Nat Commun, 2015, 6: 8132.

513. GAO S, ZHENG C, CHANG G, et al. Unique features of mutations revealed by sequentially reprogrammed induced pluripotent stem cells. Nat Commun, 2015, 18(6): 6318.

514. LIU P Y, DOU X Y, LIU C, et al. Histone deacetylation promotes mouse neural induction by restricting Nodal-dependent mesendoderm fate. Nat Commun, 2015, 6: 6830.

515. WANG H B, ZHANG H, ZHANG J P, et al. Neuropilin 1 is an entry factor that promotes EBV infection of nasopharyngeal epithelial cells. Nat Commun, 2015, 6: 6240.

516. HU Y F, CHEN Z H, FU Y J, et al. The amino-terminal structure of human fragile X mental retardation protein obtained using precipitant-immobilized imprinted polymers. Nat Commun, 2015, 6: 6634.

517. ZHU P, WANG Y, WU J, et al. LncBRM initiates YAP1 signalling activation to drive self-renewal of liver cancer stem cells. Nat Commun, 2016, 7: 13608.

518. WANG Y, WU Q, YANG P, et al. LSD1 co-repressor Rcor2 orchestrates neurogenesis in the developing mouse brain. Nat Commun, 2016, 22(7): 10481.

519. FANG J, CHENG J, WANG J, et al. Hemi-methylated DNA Opens a Closed Conformation of UHRF1 to Facilitate Its Histone Recognition. Nat Commun, 2016, 7: 11197.

520. ZHAO Q, ZHANG J Q, CHEN R Y, et al. Dissecting the precise role of H3K9 methylation in crosstalk with DNA maintenance methylation in mammals. Nat Commun, 2016, 7: 12464.

521. LIU Z, YAO X, YAN G, et al. Mediator MED23 cooperates with RUNX2 to drive osteoblast differentiation and bone development. Nat Commun, 2016, 1(7): 11149.

522. DUAN Y, HUO D W, GAO J, et al. Ubiquitin ligase RNF20/40 facilitates spindle assembly and promotes breast carcinogenesis through stabilizing motor protein Eg5. Nat Commun, 2016, 7: 12648.

523. LI L, SHI L, YANG S, et al. SIRT7 is a histone desuccinylase that functionally links to chromatin compaction and genome stability. Nat Commun, 2016, 7: 12235.

524. LI X, SONG N, LIU L, et al. USP9X regulates centrosome duplication and promotes breast carcinogenesis. Nat Commun, 2017, 31: 14866.

525. QI Z, LI Y, ZHAO B, et al. BMP restricts stemness of intestinal Lgr5+ stem cells by directly suppressing their signature genes. Nat Commun, 2017, 8: 13824.

526. LIU C, LIU W, YE Y, et al. Ufd2p synthesizes branched ubiquitin chains to promote the degradation of substrates modified with atypical chains. Nat Commun, 2017, 8: 14274.

527. ZHANG Y, ZHANG D, LI Q, et al. Nucleation of DNA repair factors by FOXA1 links DNA demethylation to transcriptional pioneering. Nat Genet, 2016, 48: 1003-1013.

528. DU C, LIU C, KANG J, et al. MicroRNA miR-326 regulates T(H)-17 differentiation and is associated with the pathogenesis of multiple sclerosis. Nat Immunol, 2009, 10: 1252-1259.

529. WANG J, LI B X, GE P P, et al. Mycobacterium tuberculosis suppresses innate immunity by coopting the host ubiquitin system. Nat Immunol, 2015, 16: 237-245.

530. XIA P, YE B, WANG S, et al. Glutamylation of the DNA sensor cGAS regulates its binding and synthase activity in antiviral immunity. Nat Immunol, 2016, 17(4): 369-378.

531. XIA P, WANG S, YE B, et al. Sox2 functions as a sequence-specific DNA sensor in neutrophils to initiate innate immunity against microbial infection. Nat Immunol, 2015, 16(4): 366-375.

532. LIU C P, XIONG C, WANG M, et al. Structure of the variant histone H3. 3. H4 heterodimer in complex with its chaperone DAXX. Nat Struct Mol Biol, 2012, 19: 1287-1293.

533. YANG D, FANG Q, WANG M, et al. N-acetylated Sir3 stabilizes the conformation of a nucleosome. binding loop in the BAH domain. Nat Struct Mol Biol, 2013, 20: 116-118.

534. ZHU P, WANG Y, HUANG G, et al. lnc-β-Catm elicits EZH2-dependent β-catenin stabilization and sustains liver CSC self-renewal. Nat Struct Mol Biol, 2016, 23(7): 631-639.

535. YU C, JI SY, SHA QQ, et al. BTG4 is a meiotic cell cycle–coupled maternal-zygotic-transition licensing factor in oocytes. Nat Struct Mol Biol, 2016, 23: 387-394.

536. HU S, WANG X, SHAN G. Insertion of an Alu element in a lncRNA leads to primate-specific modulation of alternative splicing. Nat Struct Mol Biol, 2016, 23(11): 1011-1019.

537. HUANG P, HE Z, JI S, et al. Induction of functional hepatocyte-like cells from mouse fibroblasts by defined factors. Nature, 2011, 475(7356): 386-389.

538. LI W, SHUAI L, WAN H, et al. Androgenetic haploid embryonic stem cells produce live transgenic mice. Nature, 2012, 490(7420): 407-411.

539. ROADMAP E C, ANSHUL K, WOUTER M, et al. Integrative analysis of 111 reference human epigenomes. Nature, 2015, 518(7539): 317-330.

540. DANNY L, INKYUNG J, NISHA R, et al. Integrative analysis of haplotype-resolved epigenomes across human tissues. Nature, 2015, 518(7539): 350-364.

541. HU L, LU J, CHENG J, et al. Structural insight into substrate preference for TET-mediated oxidation. Nature, 2015, 527: 118-122.

542. GUO X, WANG L, LI J, et al. Structural insight into autoinhibition and histone H3-induced activation of DNMT3A. Nature, 2015, 517: 640-644.

543. ZHANG B J, ZHENG H, HUANG B, et al. Allelic reprogramming of the histone modification H3K4me3 in early mammalian development. Nature, 2016, 537(7621):

553-557.

544. LIU X, WANG C, LIU W, et al. Distinct features of H3K4me3 and H3K27me3 chromatin domains in pre-implantation embryos. Nature, 2016, 37(7621): 558-562.

545. CHEN J, SUO S B, TAM P P, et al. Spatial transcriptomic analysis of cryosectioned tissue samples with Geo-seq. Nat Protoc, 2017, 12: 566-580.

546. YU M, SUO H, LIU M, et al. NRSF/REST neuronal deficient mice are more vulnerable to the neurotoxin MPTP. Neurobiol Aging, 2013, 34: 916-927.

547. ZHANG Q, XUE P, LI H, et al. Histone modification mapping in human brain reveals aberrant expression of histone H3 lysine 79 dimethylation in neural tube defects. Neurobiol Dis, 2013, 54: 404-413.

548. WANG Q, ZHAO X, HE S, et al. Differential proteomics analysis of specific carbonylated proteins in the temporal cortex of aged rats: the deterioration of antioxidant system. Neurochem Res, 2010, 35(1): 13-21.

549. SHANG Q, ZHOU C, LIU D, et al. Association between osteopontin gene polymorphisms and cerebral palsy in a Chinese population. Neuromolecular Med, 2016, 18(2): 232-238.

550. QI Y, WANG J, VALERIE C B, et al. Hyper-SUMOylation of the Kv7 Potassium Channel Diminishes the M-Current Leading to Seizures and Sudden Death. Neuron, 2014, 83(5): 1159-1171.

551. HE M, LIU Y, WANG X, et al. Cell-type based analysis of microRNA profiles in the mouse brain. Neuron, 2012, 73(1): 35-48.

552. YU M, CAI L, LIANG M, et al. Alteration of NRSF expression exacerbating 1-methyl-4-phenyl-pyridinium ion-induced cell death of SH-SY5Y cells. Neurosci Res, 2009, 65: 256-264.

553. CAI L, BIAN M, LIU M, et al. Ethanol-induced neurodegeneration in nrsf/rest neuronal conditional knockout mice. Neuroscience, 2011, 181: 196-205.

554. WAN P, ZHANG Y P, YAN J, et al. Glutamate enhances the surface distribution and release of Munc18 in cerebral cortical neurons. Neurosci Bull, 2010, 26: 273-281.

555. ZHANG Y P, WAN P, WANG H Q, et al. Effect of neuronal excitotoxicity on Munc18-1 distribution in nuclei of rat hippocampal neuron and primary cultured neuron. Neurosci Bull, 2011, 27: 163-172.

556. XU Y X, WANG H Q, YAN J, et al. Antibody binding to cell surface amyloid precursor protein induces neuronal injury by deregulating the phosphorylation of focal adhesion signaling related proteins. Neurosci Letters, 2009, 465: 276-281.

557. WANG H Q, XU Y X, ZHU C Q. Upregulation of heme oxygenase-1 by acteoside through ERK and PI3 K/Akt pathway confer neuroprotection against beta-amyloid-

induced neurotoxicity. Neurotox Res, 2012, 21: 368-378.
558. SHEN L, GAO G, ZHANG Y, et al. A single amino acid substitution confers enhanced methylation activity of mammalian Dnmt3b on chromatin DNA. Nucl Acids Res, 2010, 38(18): 6054-6064.
559. REN J, LIU Z, GAO X, et al. MiCroKit 3.0: an integrated database of midbody, centrosome and kinetochore. Nucl Acids Res, 2010, 38: D155-D160.
560. HE J, YE J, CAI Y, et al. Structure of p300 bound to MEF2 on DNA reveals a mechanism of enhanceosome assembly. Nucl Acids Res, 2011, 39(10): 4464-4474.
561. LIU Z, CAO J, GAO X, et al. CPLA 1. 0: an integrated database of protein lysine acetylation. Nucl Acids Res, 2011, 39: D1029-D1034.
562. PENG J, ZHOU J Q. The tail-module of yeast mediator complex is required for telomere heterochromatin maintenance. Nucl Acids Res, 2012, 40(2): 581-593.
563. SHI J, ZHENG M, YE Y, et al. Drosophila Brahma complex remodels nucleosome organizations in multiple aspects. Nucl Acids Res, 2014, 42: 9730-9739.
564. HE C, WANG X W, MICHAEL Q Z. Nucleosome eviction and multiple co-factor binding predict estrogen-receptor-alpha-associated long-range interactions. Nucl Acids Res, 2014, 42(11): 6935-6944.
565. DENG Z Q, KATHLEEN C L, LI X R, et al. Structural basis for the regulatory function of a complex zinc-binding domain in a replicative arterivirus helicase resembling a nonsense-mediated mRNA decay helicase. Nucl Acids Res, 2014, 42(5): 3464-3477.
566. GUO W L, CHUNG W Y, QIAN M P, et al. Characterizing the strand-specific distribution of non-CpG methylation in human pluripotent cells. Nucl Acids Res, 2014, 42(5): 3009-3016.
567. FENG Z X, LI J, ZHANG J R, et al. qDNAmod: a statistical model-based tool to reveal intercellular heterogeneity of DNA modification from SMRT sequencing data. Nucl Acids Res, 2014, 42(22): 13488-13499.
568. CHEN Y, DING Y F, YANG L, et al. Integrated omics study delineates the dynamics of lipid droplets in rhodococcus opacus PD630. Nucl Acids Res, 2014, 42(2): 1052-1064.
569. WEI T, CHEN W, WANG X, et al. An HDAC2-TET1 switch at distinct chromatin regions significantly promotes the maturation of pre-iPS to iPS cells. Nucl Acids Res, 2015, 43(11): 5409-5422.
570. ZOU D, SUN S X, LI R J, et al. MethBank: a database integrating next-generation sequencing single-base-resolution DNA methylation programming data. Nucl Acids Res, 2015, 43: 54-58.

571. HU L, DI C, KAI M X, et al. A common set of distinct features that characterize noncoding RNAs across multiple species. Nucl Acids Res, 2015, 43(1): 104-114.

572. LIU X, CHEN Z, XU C X, et al. Repression of hypoxia-inducible factor a signaling by set7-mediated methylation. Nucl Acids Res, 2015, 43(10): 5081-5098.

573. XU Z, SONG Z, LI G, et al. H2B ubiquitination regulates meiotic recombination by promoting chromatin relaxation. Nucl Acids Res, 2016, 44(20): 9681-9697.

574. CHEN Y, WANG Y F, XUAN Z Y, et al. De novo deciphering three-dimensional chromatin interaction and topological domains by wavelet transformation of epigenetic profiles. Nucl Acids Res, 2016, 44(11): e106.

575. MA X P, ZHAN G, MONICA C S, et al. Analys is of C. elegans muscle transcriptome using trans-splicing-based RNA tagging (SRT). Nucl Acids Res, 2016, 44(21): 734.

576. CHEN S, JING Y, KANG X, et al. Histone HB mono-ubiquitination is a critical epigenetic switch for the regulation of autophagy. Nucl Acids Res, 2017, 45 (3): 1144-1158.

577. TU Y F, LIU H M, ZHU X F, et al. Ataxin-3 promotes genome integrity by stabilizing Chk1. Nucl Acids Res, 2017, 45(8): 4532.

578. LI G P, CHEN Y, MICHAEL P S, et al. ChIA-PET2: a versatile and flexible pipeline for ChIA-PET data analysis. Nucl Acids Res, 2017, 45(1): 809.

579. CHEN P, WANG Y, LI G H. Dynamics of histone variant H3. 3 and its coregulation with H2A. Z at enhancers and promoters. Nucleus, 2014, 5: 21-27.

580. LIU Y, FANG Y. Nucleolus-tethering System (NoTS): an intranuclear fluorescence two-hybrid assay for studying the nuclear protein-protein interactions and nuclear body initiation in vivo. Nucleus, 2014, 5: 287-293.

581. YU B, SHAN G. Functions of long noncoding RNAs in the nucleus. Nucleus, 2016, 7(2): 155-166.

582. GAO X N, LIN J, LI Y H, et al. MicroRNA-193a represses c-kit expression and functions as a methylation-silenced tumor suppressor in acute myeloid leukemia. Oncogene, 2011, 30: 3416-3428.

583. DOU L, ZHENG D, LI J, et al. Methylation-mediated repression of microRNA-143 enhances MLL-AF4 oncogene expression. Oncogene, 2012, 31: 507-517.

584. JIN W, WU K, LI Y Z, et al. AML1-ETO targets and suppresses cathepsin G, a serine protease, which is able to degrade AML1-ETO in t(8;21) acute myeloid leukemia. Oncogene, 2013, 32: 1978-1987.

585. YANG S, LI Y, GAO J, et al. MicroRNA-34 suppresses breast cancer invasion and metastasis by directly targeting Fra-1. Oncogene, 2013, 32: 4294-4303.

586. SHAN L, LI X, LIU L, et al. GATA3 Cooperates with PARP1 to Regulate CCND1 Transcription through Modulating Histone H1 Incorporation. Oncogene, 2014, 33: 3205-3216.
587. XU C, FU H, GAO L, et al. BCR-ABL/GATA1/miR-138 mini circuitry contributes to the leukemogenesis of chronic myeloid leukemia. Oncogene, 2014, 33: 44-54.
588. FAN D, ZHOU X, LI Z, et al. Stem cell programs are retained in human leukemic lymphoblasts. Oncogene, 2015, 34: 2083-2093.
589. CAO L L, WEI F, DU Y, et al. ATM-mediated KDM2A phosphorylation is required for the DNA damage repair. Oncogene, 2016, 35(3): 301-313.
590. HE X H, ZHU W, YUAN P, et al. miR-155 downregulates ErbB2 and suppresses ErbB2-induced malignant transformation of breast epithelial cells. Oncogene, 2016, 35(46): 6015-6025.
591. ZHANG P, HE X, TAN J, et al. β-arrestin2 mediates beta-2 adrenergic receptor signaling prostate cancer cell progression. Oncol Rep, 2011, 26: 1471-1477.
592. XIONG D, YE Y L, CHEN M K, et al. Non-muscle myosin II is an independent predictor of overall survival for cystectomy candidates with early-stage bladder cancer. Oncol Rep, 2012, 28: 1625-1632.
593. ZHANG P, CHEN Y, JIANG X, et al. Tumor-targeted efficiency of shRNA vector harboring chimera hTERT/U6 promoter. Oncol Rep, 2010, 23: 1309-1316.
594. YAO Y, LI C, ZHOU X, et al. PIWIL2 Induces c-Myc expression by interacting with NME2 and regulates c-Myc-mediated tumor cell proliferation. Oncotarget, 2014, 5(18): 8466-8477.
595. YAO L, REN S, ZHANG M, et al. Identification of Specific DNA Methylation Sites on the Y-chromosome as biomarker in Prostate Cancer. Oncotarget, 2015, 6(38): 40611-40621.
596. LU S, YANG Y, DU Y, et al. The transcription factor c-Fos coordinates with histone lysine-specific demethylase 2A to activate the expression of cyclooxygenase-2. Oncotarget, 2015, 6(33): 34704-34717.
597. LI C, ZHOU X, CHEN J, et al. PIWIL1 destabilizes microtubule by suppressing phosphorylation at Ser16 and RLIM-mediated degradation of stathmin1. Oncotarget, 2015, 6(29): 27794-27804.
598. LI X, XU X, FANG J, et al. Rs2853677 modulates Snail1 binding to the TERT enhancer and affects lung adenocarcinoma susceptibility. Oncotarget, 2016, 7(25): 37825-37838.
599. YANG Y, HUANG Y, WANG Z, et al. HDAC10 promotes lung cancer proliferation via AKT phosphorylation. Oncotarget, 2016, 7(37): 59388-59401.

600. DUAN Y, WU X, ZHAO Q, et al. DOT1L promotes angiogenesis through cooperative regulation of VEGFR2 with ETS-1. Oncotarget, 2016, 7(43): 69674-69687.
601. SUN X, LIU S, CHEN P, et al. miR-449a inhibits colorectal cancer progression by targeting SATB2. Oncotarget, 2016, 8(60): 100975-100988.
602. ZHU X F, MA X L, TU Y F, et al. Parkin regulates translesion DNA synthesis in response to UV radiation. Oncotarget, 2017, 8(22): 36423-36437.
603. LI C, MA Y, ZHANG K S, et al. Aberrant transcriptional networks in step-wise neurogenesis of paroxysmal kinesigenic dyskinesia-induced pluripotent stem cells. Oncotarget, 2016, 7: 33.
604. ZHANG L, WANG G, WANG L, et al. Valproic acid inhibits prostate cancer cell migration by up-regulating E-cadherin expression. Pharmazie, 2011, 66: 614-618.
605. LU J, HU L, CHENG J, et al. A computational investigation on the substrate preference of ten-eleven-translocation 2 (TET2). Phys Chem Chem Phys, 2016, 18(6): 4728-4738.
606. CI W M, LIU J. Programming and inheritance of parental DNA methylomes in vertebrates. Physiology, 2015, 30: 63-68.
607. LI G, ZHANG J, LI J, et al. Imitation switch chromatin remodeling factors and their interacting RINGLET proteins act together in controlling the plant vegetative phase in Arabidopsis. Plant J, 2012, 72: 261-270.
608. ZHANG T W, ZHANG X W, HU S N, et al. An efficient procedure for organellar genome assembly using whole genome data from 454 GS FLX sequencing. Plant Methods, 2011, 7: 38.
609. LIU Q, SHI L L, FANG Y. Dicing bodies. Plant Physiol, 2012, 158: 61-66.
610. LIU Q, YAN Q Q, LIU Y, et al. Complementation of hyponastic leaves1 by double-strand RNA binding domains of dicer-like 1 in nuclear dicing bodies. Plant Physiol, 2013, 163: 108-117.
611. HU X, XU L. Transcription factors WOX11/12 directly activate WOX5/7 to promote root primordia initiation and organogenesis. Plant Physiol, 2016, 172: 2363-2373.
612. CHEN X, CHENG J, CHEN L, et al. Auxin-independent NAC pathway acts in response to explant-specific wounding and promotes root tip emergence during de novo root organogenesis in Arabidopsis. Plant Physiol, 2016, 170: 2136-2145.
613. CHEN L, SUN B, XU L, et al. Wound signaling: the missing link in plant regeneration. Plant Sig Behav, 2016, 11: e1238548.
614. LI Y, ZHU J, TIAN G. The DNA methylome of human peripheral blood mononuclear cells. PLoS Biol, 2010, 8: e1000533.

615. HUANG S J, ZHANG Z, ZHANG C X, et al. Activation of smurf E3 ligase promoted by smoothened regulates hedgehog signaling through targeting patched turnover. PLoS Biol, 2013, 11: e1001721.
616. WANG F, HE L, HUANGYANG P W, et al. JMJD6 promotes colon carcinogenesis through negative regulation of p53 by hydroxylation. PLoS Biol, 2014, 12(3): e1001819.
617. LIU H, LI X, NING G, et al. The Machado–joseph disease deubiquitinase Ataxin-3 regulates the stability and apoptotic function of p53. PLoS Biol, 2016, 14(11): e2000733.
618. HUANG J L, NIU C Q, CHRISTOPHER D G, et al. Systematic prediction of pharmacodynamic drug-drug interactions through protein-protein-interaction network. PLoS Comput Biol, 2013, 9(3): e1002998.
619. ZHOU B O, WANG S S, ZHANG Y, et al. Histone H4 lysine 12 acetylation regulates telomeric heterochromatin plasticity in saccharomyces cerevisiae. PLoS Genetics, 2011, 7(1): e1001272.
620. HE C, CHEN X, HUANG H, et al. Reprogramming of H3K27me3 is critical for acquisition of pluripotency from cultured Arabidopsis tissues. PLoS Genetics, 2012, 8: e1002911.
621. LI J, LIU Y, LIU M, et al. Functional dissection of regulatory models using gene expression data of deletion mutants. PLoS Genetics, 2013, 9(9): e1003757.
622. KIRAN Z, ZHAO J H, NEIL A S, et al. Nicotiana small RNA sequences support a host genome origin of Cucumber Mosaic Virus Satellite RNA. PLoS Genetics, 2015, 11: e1004906.
623. SUN Z, GUO T T, LIU Y, et al. The Roles of arabidopsis CDF2 in transcriptional and posttranscriptional regulation of primary microRNAs. PLoS Genetics, 2015, 11: e1005598.
624. WANG H, LIU C, CHENG J, et al. Arabidopsis flower and embryo development are repressed in seedlings by different combinations of Polycomb group proteins in association with distinct sets of cis-regulatory elements. PLoS Genetics, 2016, 12: e1005771.
625. YAN Q, XIA X, SUN Z, et al. Depletion of arabidopsis SC35 and SC35-like serine/arginine-rich proteins affects the transcription and splicing of a subset of genes. PLoS Genetics, 2017, 13(3): e1006663.
626. KONG X, WANG R, XUE Y, et al. Sirtuin 3, a new target of PGC-1a, plays an important role in the suppression of ROS and mitochondrial biogenesis. PLoS One, 2010, 5: e11707.

627. LI Q, WANG X, LU Z M, et al. Polycomb CBX7 directly controls trimethylation of histone H3 at lysine 9 at the p16 locus. PLoS One, 2010, 5: e13732.
628. XUE Y, LIU Z, GAO X, et al. GPS-SNO: Computational prediction of protein S-nitrosylation sites with a modified GPS algorithm. PLoS One, 2010, 5: e11290.
629. SHI J T, YANG W T, CHEN M J, et al. AMD, an automated motif discovery tool using stepwise refinement of gapped consensuses. PLoS One, 2011, 6(9): e24576.
630. KONG C, WANG C, LU J, et al. NEDD9 is a positive regulator of epithelial-mesenchymal transition and promotes invasion in aggressive breast cancer. PLoS One, 2011, 6(7): e22666.
631. LI J, SHI Y, SUN J, et al. Xenopus reduced folate carrier regulates neural crest development epigenetically. PLoS One, 2011, 6: e27198.
632. XIONG J, WANG H, GUO G, et al. Male germ cell apoptosis and epigenetic histone modification induced by tripterygium wilfordii hookf. PLoS One, 2011, 6: e20751.
633. XIE X W, LIU J X, HU B, et al. Zebrafish foxo3b negatively regulates canonical wnt signaling to affect early embryogenesis. Plos One, 2011, 6(9): e24469.
634. LIU Z, CAO J, GAO X, et al. GPS-CCD: a novel computational program for the prediction of calpain cleavage sites. PLoS One, 2011, 6: e19001.
635. ZHAO H Y, ZHANG Y J, DAI H, et al. CARM1 mediates modulation of Sox2. PLoS One, 2011, 6: e277026.
636. MA L, HUANG Y, ZHU W, et al. An integrated analysis of miRNA and mRNA expressions in non-small cell lung cancers. PLoS One, 2011, 6(10): e26502.
637. CUI P, LIN Q, ZHANG L, et al. The disequilibrium of nucleosomes distribution along chromosomes plays a functional and evolutionarily role in regulating gene expression. PLoS One, 2011, 6(8): e23219.
638. GUO T, WANG W P, ZHANG H, et al. ISL1 promotes pancreatic islet cell proliferation. PLoS One, 2011, 6(8): e22387.
639. WU J, CUI N, WANG R, et al. A role for CARM1-mediated histone H3 methylation in protecting histone acetylation by releasing corepressors from chromatin. PLoS One, 2012, 7(6): e34692.
640. LIU T R, XU L H, YANG A K, et al. Decreased expression of SATB2: a novel independent prognostic marker of worse outcome in laryngeal carcinoma patients. PLoS One, 2012, 7: e40704.
641. WU K, DONG D D, FANG H, et al. An interferon-related signature in the transcriptional core response of human macrophages to mycobacterium tuberculosis infection. PLoS One, 2012, 7: e38367.
642. CHEN J, WANG G, LU C, et al. Synergetic cooperation of microRNAs with

transcription factors in iPS cell generation. PLoS One, 2012, 7(7): e40849.

643. WANG L, WANG G, LU C, et al. Contribution of the -160C/A polymorphism in E-cadherin promoter to cancer risk: a meta-analysis of 47 case-control studies. Plos One, 2012, 7(7): e40219.

644. WU G, ZHU J, HE F H, et al. Gene and genome parameters of mammalian liver circadian genes (LCG). PLoS One, 2012, 7(10): e46961.

645. LU Z M, LI Q, ZHOU J, et al. Nucleosomes correlate with in vivo progression pattern of de novo methylation of p16 CpG islands in human gastric carcinogenesis. PLoS One, 2012, 7: e35928.

646. LIU Z, REN G, SHANG G C, et al. ATRA inhibits the proliferation of DU145 prostate cancer cells through reducing the methylation level of HOXB13 gene. PLoS One, 2012, 7(7): e40943.

647. BIAN M, LIU J, HONG X, et al. Overexpression of parkin ameliorates dopaminergic neurodegeneration induced by 1-methyl-4-phenyl-1, 2, 3, 6- tetrahydropyridine in mice. PLoS One, 2012, 7: e39953.

648. LU Y L, ZHANG K, LI C, et al. Piwil2 suppresses P53 by inducing phosphorylation of signal transducer and activator of transcription 3 in tumor cells. PLoS One, 2012, 7(1): e30999.

649. ZHANG K, LU Y L, YANG P, et al. HILI inhibits TGF-β signaling by interacting with Hsp90 and promoting TβR degradation. PLoS One, 2012, 7(7): e41973.

650. LIU Z, YUAN F, REN J, et al. GPS-ARM: computational analysis of the APC/C recognition motif by predicting D-boxes and KEN-boxes. PLoS One, 2012, 7: e34370.

651. CAI R, LIU Z, REN J, et al. GPS-MBA: computational analysis of MHC class II epitopes in type 1 diabetes. PLoS One, 2012, 7: e33884.

652. LIU K, BIAN C, LIU H, et al. Crystal structure of TDRD3 and methyl. arginine binding characterizations of TDRD3, SMN and SPF30. PLoS One, 2012, 7: e30375.

653. ZHANG T W, FAN Y J, DENG X, et al. The complete chloroplast and mitochondrial genome sequences of resurrection plant boea hygrometrica: insights into the evolution of plant organellar genomes. PLoS One, 2012, 7(1): e30531.

654. WANG Y, LI H S, WANG X, et al. Hypoxia promotes dopaminergic differentiation of mesenchymal stem cells and shows benefits for transplantation in a rat model of parkinson's disease. PLoS One, 2013, 8(1): e54296.

655. QIAO N, HUANG Y, HAMMAD N, et al. Cociter: an efficient tool to infer gene function by assessing the significance of literature co-citation. PLoS One, 2013, 8(9): e74074.

656. CHEN G, SHI X, SUN C, et al. VEGF-mediated proliferation of human adipose tissue-derived stem cells. PLoS One, 2013, 8: e73673.

657. SHI J, HU J, ZHOU Q, et al. PEpiD: a prostate epigenetic database in mammals. PLoS One, 2013, 8: e64289.

658. TANG Y, DONG S, CAO X, et al. H2A. Z Nucleosome positioning has no impact on genetic variation in drosophila genome. PLoS One, 2013, 8: e58295.

659. LIU Y, LIU Q, JIA W, et al. MicroRNA-200a regulates Grb2 and suppresses differentiation of mouse embryonic stem cells into endoderm and mesoderm. PLoS One, 2013, 8(7): e68990.

660. SI X, CHEN W, GUO X, et al. Activation of GSK3b by sirt2 is required for early lineage commitment of mouse embryonic stem cell. PLoS One, 2013, 8(10): e76699.

661. YUE M, LI Q, ZHANG Y, et al. Histone H4R3 methylation catalyzed by SKB1/PRMT5 is required for maintaining shoot apical meristem. PLoS One, 2013, 8: e83528.

662. YANG C, GU L, DENG D. Bone marrow-derived cells may not be the original cells for carcinogen-induced mouse gastrointestinal carcinomas. PLoS One, 2013, 8: e079615.

663. ZHANG C, HONG Z, MA W, et al. Drosophila UTX coordinates with p53 to regulate ku80 expression in response to dna damage. PLoS One, 2013, 8(11): e78652.

664. HONG Z, JIANG J, MA J, et al. The Role of hnRPUL1 involved in dna damage response is related to PARP1. PLoS One, 2013, 8(4): e60208.

665. GAO X N, LIN J, NING Q Y, et al. A histone acetyltransferase p300 inhibitor C646 induces cell cycle arrest and apoptosis selectively in AML1-ETO-positive AML cells. PLoS One, 2013, 8: e55481.

666. ZHANG X L, JIANG L, WANG G Q, et al. Structural insights into the abscisic acid stereospecificity by the ABA receptors PYR/PYL/RCAR. PLoS One, 2013, 8: e67477.

667. WU D M, GU J, MICHAEL Q Z. FastDMA : an infinium humanmethylation450 beadchip analyzer. PLoS One 2013, 8(9): e74275.

668. CHEN P, ZHANG J, ZHAN Y, et al. Established thymic epithelial progenitor/stem cell-like cell lines differentiate into mature thymic epithelial cells and support T cell development. PLoS One, 2013, 8: e75222.

669. YIN R, GU L, LI M, et al. Gene expression profiling analysis of bisphenol a-induced perturbation in biological processes in er-negative HEK 293 cells. PLoS One, 2014, 9: e98635.

670. LV W, GUO X, WANG G, et al. Histone deacetylase 1 and 3 regulate the mesodermal lineage commitment of mouse embryonic stem cells. PLoS One, 2014, 9(11): 1-19.

671. LI G P, LI M, ZHANG Y W, et al. Module Role: a tool for modulization, role determination and visualization in protein-protein interaction networks. PLoS One, 2014, 9(5): e94608.

672. FAN J D, LEI P J, ZHENG J Y, et al. The selective activation of P53 target genes regulated by Smyd2 in Bix-01294 induced autophagy-related cell death. PLoS One, 2015, 10: e0116782.

673. ZHANG M, WANG C, YANG C, et al. Epigenetic pattern on the human Y chromosome is evolutionarily conserved. PLoS One, 2016, 11(1): e0146402.

674. LIANG L, SUN H, ZHANG W, et al. Meta-analysis of EMT datasets reveals different types of EMT. PLoS One, 2016, 11(6): e0156839.

675. HUA J L, ZHOU B O, ZHANG R R, et al. The N-terminus of histone H3 is required for de novo DNA methylation in chromatin. PNAS, 2009, 106(52): 22187-22192.

676. WANG G, CUI Y, ZHANG G, et al. Regulation of proto-oncogene transcription, cell proliferation, and tumorigenesis in mice by PSF protein and a VL30 noncoding RNA. PNAS, 2009, 106(39): 16794-16798.

677. LI L, FENG T, LIAN Y, et al. Role of human noncoding RNAs in the control of tumorigenesis. PNAS, 2009, 106 (31): 12956-12961.

678. SHENG N Y, XIE Z H, WANG C, et al. Retinoic acid regulates bone morphogenic protein signal duration by promoting the degradation of phosphorylated Smad1. PNAS, 2010, 107: 18886-18891.

679. XIAN D, GU L F, LIU C Y, et al. Arginine methylation mediated by the Arabidopsis homolog of PRMT5 is essential for proper pre-mRNA splicing. PNAS, 2010, 107: 19114-19119.

680. XIE Z H, CHEN Y F, LI Z F, et al. Smad6 promotes neuronal differentiation in the intermediate zone of the dorsal neural tube by inhibition of the Wnt/β-catenin pathway. PNAS, 2011, 108: 12119-12124.

681. LIU X, WANG D, ZHAO Y, et al. Methyltransferase Set7/9 regulates p53 activity by interacting with sirt1. PNAS, 2011, 108(5): 1925-1930.

682. BIAN X, ROBIN W K, TINA Y L, et al. Structures of the atlastin GTPase provide insight into homotypic fusion of endoplasmic reticulum membranes. PNAS, 2011, 108: 3976-3981.

683. SHI L L, WANG J, HONG F, et al. Four amino acids guide the assembly or

disassembly of Arabidopsis histone H3. 3-containing nucleosomes. PNAS, 2011, 108: 10574-10578.
684. SUN L, WANG M, LV Z, et al. Structural insights into protein arginine symmetric dimethylation by PRMT5. PNAS, 2011, 108: 20538-20543.
685. ZHOU B, YANG L, LI S F, et al. Midlife gene expressions identify modulators of aging through dietary interventions. PNAS, 2012, 109(19): e1201-e1209.
686. AGUSTIN C, AVNISH K, WANG X W, et al. H3K4 demethylation by Jarid1a and Jarid1b contributes to retinoblastoma-mediated gene silencing during cellular senescence. PNAS, 2012, 109(23): 8971-8976.
687. YANG N, WANG W, WANG Y, et al. Distinct mode of methylated lysine-4 of histone H3 recognition by tandem tudor-like domains of Spindlin1. PNAS, 2012, 109: 17954-17959.
688. WANG D L, ZHOU J Y, LIU X Y, et al. Methylation of SUV39H1 by SET7/9 results in heterochromatin relaxation and genome instability. PNAS, 2013, 110(14): 5516-5521.
689. WANG G, GUO X, HONG W, et al. Critical regulation of miR-200/ZEB2 pathway in Oct4/Sox2-induced mesenchymal-to-epithelial transition and induced pluripotent stem cell generation. PNAS, 2013, 110(8): 2858-2863.
690. ZHANG Y, CHEN L Y, HAN X, et al. Phosphorylation of TPP1 regulates cell cycle-dependent telomerase recruitment. PNAS, 2013, 110(14): 5457-5462.
691. ZHANG H, MI J Q, FANG H, et al. Preferential eradication of acute myelogenous leukemia stem cells by fenretinide. PNAS, 2013, 110: 5606-5611.
692. LI X, LIU L, YANG S D, et al. Histone demethylase KDM5B is a key regulator of genome stability. PNAS, 2014, 111: 7096-7101.
693. ZHANG P, TU B, WANG H, et al. Tumor suppressor p53 cooperates with SIRT6 to regulate gluconeogenesis by promoting FoxO1 nuclear exclusion. PNAS, 2014, 111(29): 10684-10689.
694. DUAN H, GE W, ZHANG A, et al. Transcriptome analyses reveal molecular mechanisms underlying functionalrecovery after spinal cord injury. PNAS, 2015, 112(43): 13360-13365.
695. YANG Z Y, ZHANG A F, DUAN H M, et al. NT3-chitosan elicits robust endogenous neurogenesis to enable functional recovery after spinal cord injury. PNAS, 2015, 112(43): 13354-13359.
696. YUAN Y, LIU B, XIE P, et al. Model-guided quantitative analysis of microRNA-mediated regulation on competing endogenous RNAs using a synthetic gene circuit. PNAS, 2015, 112(10): 3158-3163.

697. PAN L, XIE W, LI K, et al. Heterochromatin remodeling by CDK12 contributes to learning. PNAS, 2015, 112(45): 13988-13993.
698. CHENG M Y, LI X R, RUI M, et al. ATP binding by the P-loop NTPase OsYchF1 (an unconventional G protein) contributes to biotic but not abiotic stress responses. PNAS, 2016, 113: 2648-2653.
699. YE C, ZHANG X G. A simulation study on gene expression regulation via stochastic model. Proceedings of the 33rd Chinese Control Conference, 2014, 6885-6888.
700. WU F Y, FENG Q, HENG M, et al. The activation of excitatory amino acid receptors is involved in tau phosphorylation induced by cold water stress. Prog Biochem Biophys, 2010, 37: 510-516.
701. XU Y X, WANG H Q, ZHAO H, et al. Intrahippocampus injection of antibodies to amyloid beta-protein precursor causes cognitive deficits and neuronal degeneration. Prog Biochem Biophys, 2011, 38: 908-918.
702. WANG P, CHEN L P, LI Q Q, et al. Cellular senescence and senescent cell. Prog Biochem Biophys, 2012, 39: 257-263.
703. GU Q, HAO J, ZHAO X Y, et al. Rapid conversion of human ESCs into mouse ESC-like pluripotent state by optimizing culture conditions. Protein & Cell, 2012, 3(1): 71-79.
704. NIU L F, LU F L, ZHAO T L, et al. The enzymatic activity of Arabidopsis protein arginine methyltransferase 10 is essential for flowering time regulation. Protein & Cell, 2012, 3: 450-459.
705. CHEN X, WEI S, YANG F. Mitochondria in the pathogenesis of diabetes: a proteomic view. Protein & Cell, 2012, 3(9): 648-660.
706. SONG Y R, HAI T, WANG Y, et al. Epigenetic reprogramming, gene expression and in vitrodevelopment of porcine SCNT embryos are significantly improved by a histone deacetylase inhibitor—m-carboxycinnamic acid bishydroxamide (CBHA). Protein & Cell, 2014, 5(5): 382-393.
707. CHEN P, ZHU P, LI G H. New insights into the helical structure of 30-nm chromatin fibers. Protein & Cell, 2014, 7: 489-491.
708. MENG J, ZOU X, WU R, et al. Accelerated regeneration of skeletal muscle in rnf13-knockout mice is mediated by macrophage-secreted IL-4/IL-6. Protein & Cell, 2014, 5(3): 235-247.
709. HUANG Y, LIANG P, LIU D, et al. Telomere regulation in pluripotent stem cells. Protein & Cell, 2014, 5(3): 194-202.
710. LIANG P, XU Y, ZHANG X, et al. CRISPR/Cas9-mediated gene editing in human

tripronuclear zygotes. Protein & Cell, 2015, 6(5): 363-372.

711. XU M, YANG X, YANG X A, et al. Structural insights into the regulatory mechanism of the Pseudomonas aeruginosa. Protein & Cell, 2016, 7(6): 403-416.

712. YANG H, WANG J, LIU M, et al. A resolution cryo-EM structure of human mTOR Complex 1. Protein & Cell, 2016, 7(12): 878-887.

713. CHEN X Y, ZHANG K S, ZHOU L Q, et al. Coupled electrophysiological recording and single cell transcriptome analyses revealed molecular mechanisms underlying neuronal maturation. Protein & Cell, 2016, 7(3): 175-186.

714. ZHAO D, ZHANG X J, GUAN H P, et al. The BAH domain of BAHD1 is a histone H3K27me3 reader. Protein & Cell, 2016, 7(3): 222-226.

715. XUE Y, LIU Z, CAO J, et al. GPS 2.1: enhanced prediction of kinase-specific phosphorylation sites with an algorithm of motif length selection. Protein Engineering Design & Selection, 2011, 24: 255-260.

716. LIU Y, WU J, YAN G, et al. Proteomic analysis of prolactinoma cells by immuno-laser capture microdissection combined with online two-dimensional nano-scale liquid chromatography/mass spectrometry. Proteome Sci, 2010, 8(1): 1-11.

717. REN J, GAO X, JIN C, et al. Systematic study of protein sumoylation: development of a site-specific predictor of SUMOsp 2.0. Proteomics, 2009, 9: 3409-3412.

718. ZHANG K, CHEN Y, ZHANG Z, et al. Unrestrictive identification of non-phosphorylation PTMs in yeast kinases by MS and PTMap. Proteomics, 2010, 10(5): 896-903.

719. CHEN X, CUI Z, WEI S, et al. Chronic high glucose-induced INS-1 β cell mitochondrial dysfunction: a comparative mitochondrial proteome with SILAC. Proteomics, 2013, 13(20): 3030-3039.

720. HAMMAD N, HAN J D. Structure-based protein-protein interaction networks and drug design. Quant Biol, 2013, 1(3): 183-191.

721. GAO J T, YANG X S, MOHAMED N D, et al. Developing bioimaging and quantitative methods to study 3D genome. Quant Biol, 2016, 4(2): 129-147.

722. HE C, LI G P, DIEKIDEL M N, et al. Advances in computational ChIA-PET data analysis. Quant Biol, 2016, 4(3): 217-225.

723. ZHOU D, WU D M, LI Z, et al. Population dynamics of cancer cells with cell state conversions. Quant Biol, 2013, 1(3): 201-208.

724. AN M, DAI J, WANG Q, et al. Efficient and clean charge derivatization of peptides for analysis by mass spectrometry. Rapid Commun Mass Spectrom, 2010, 24(13): 1869-1874.

725. GAO H, ZHANG H L, SHOU J, et al. Towards retinal ganglion cell regeneration. Regen Med, 2012, 7: 865-875.
726. YU C, LI J, YUAN Z, et al. Two novel mutations affecting the same splice site of PKD1 correlate with different phenotypes in ADPKD. Ren Fail, 2014, 36: 687-693.
727. SHI L, WU J. Epigenetic regulation in mammalian preimplantation embryo development. Reprod Biol Endocrin, 2009, 7: 59.
728. YANG Z, SUN N, WANG S, et al. Molecular cloning and expression of a new gene, GON-SJTU1 in the rat testis. Reprod Biol Endocrin, 2010, 8: 43.
729. DENG D, LU Z M. Differentiation and adaptation epigenetic networks: translational research in gastric carcinogenesis. Sci Bull, 2013, 58: 1-6.
730. SUN B, CHEN L, LIU J, et al. TAA family contributes to auxin production during de novo regeneration of adventitious roots from Arabidopsis leaf explants. Sci Bull, 2016, 61: 1728-1731.
731. ZENG M, HU B, LI J, et al. Stem cell lineage in body layer specialization and vascular patterning of rice root and leaf. Sci Bull, 2016, 61: 847-858.
732. WANG C Z, ZHU B. You are never alone: crosstalk among epigenetic players. Sci Bull, 2016, 60(10): 899-904.
733. YAN X, PAN J, XIONG W, et al. Yin Yang 1 (YY1) synergizes with Smad7 to inhibit TGF-β signaling in the nucleus. Sci China Life Sci, 2014, 57(1): 128-136.
734. SHENG M M, ZHONG Y, CHEN Y, et al. Hsa-miR-1246, hsa-miR-320a and hsa-miR-196b-5p inhibitors can reduce the cytotoxicity of Ebola virus glycoprotein in vitro. Sci China Life Sci, 2014, 57(10): 959-972.
735. LI Y X, LIU X N, WANG X W, et al. Sequence signatures of genes with accompanying antisense transcripts in Saccharomyces cerevisiae. Sci China Life Sci, 2014, 57(1): 52-58.
736. LI M, FANG Y. Histone variants: the artists of eukaryotic chromatin. Sci China Life Sci, 2015, 58: 232-239.
737. CAO L L, SHEN C C, ZHU W G. Histone modifications in DNA damage response. Sci China Life Sci, 2016, 59(3): 257-270.
738. XIONG C Y, WEN Z Q, LI G H. Histone Variant H3. 3: A versatile H3 variant in health and in disease. Sci China Life Sci, 2016, 59: 245-256.
739. CHENG H, DOU X Y, HAN J D, et al. Understanding super-enhancers. Sci China Life Sci, 2016, 59(3): 277-280.
740. ZHANG M, ZHOU H, ZHENG C, et al. The roles of testicular c-kit positive cells in de novo morphogenesis of testis. Sci Rep, 2014, 4: 5936.

741. FU Y, LV P, YAN G, et al. MacroH2A1 associates with nuclear lamina and maintains chromatin architecture in mouse liver cells. Sci Rep, 2015, 5: 17168.
742. SUN H, LIANG L, LI Y, et al. Lysine-specific histonedemethylase 1 inhibition promotes reprogramming by facilitating the expression of exogenous transcriptional factors and metabolic switch. Sci Rep, 2016, 6: 30903.
743. YE Y, GU L, CHEN X, et al. Chromatin remodeling during the in vivo glial differentiation in early Drosophila embryos. Sci Rep, 2016, 6: 33422.
744. SHI J, HE J, LIN J, et al. Distinct response of the hepatic transcriptome to Aflatoxin B1 induced hepatocellular carcinogenesis and resistance in rats. Sci Rep, 2016, 6: 31898.
745. ZHAO J H, FANG Y Y, DUAN C G, et al. Genome-wide identification of endogenous RNA-directed DNA methylation loci associated with abundant 21-nucleotide siRNAs in Arabidopsis. Sci Rep, 2016, 6: 36247.
746. JIN W S, WANG L, ZHU F, et al. Critical POU domain residues confer Oct4 uniqueness in somatic cell reprogramming. Sci Rep, 2016, 6: 20818.
747. ZHANG L, SHAO Y, LI L, et al. Efficient liver repopulation of transplanted hepatocyte prevents cirrhosis in a rat model of hereditary tyrosinemia type I. Sci Rep, 2016, 11(6): 31460.
748. LIU L, LIU X, REN X, et al. Smad2 and Smad3 have differential sensitivity in relaying TGFbeta signaling and inversely regulate early lineage specification. Sci Rep, 2016, 6: 21602.
749. GUO W L, MICHAEL Q Z, WU H. Mammalian non-CG methylations are conserved and cell-type specific and may have been involved in the evolution of transposon elements. Sci Rep, 2016, 6: 32207.
750. WANG Y H, WANG J Q, WANG Q C, et al. Endophilin B2 promotes inner mitochondrial membrane degradation by forming heterodimers with Endophilin B1 during mitophagy. Sci Rep, 2016, 6: 25153.
751. ZHAI Z, TANG M, YANG Y, et al. Identifying human SIRT1 substrates by integrating heterogeneous information from various sources. Sci Rep, 2017, 7(1): 4614.
752. HOU P, LI Y, ZHANG X, et al. Pluripotent stem cells induced from mouse somatic cells by small-molecule compounds. Science, 2013, 341(6146): 651-654.
753. SONG F, CHEN P, SUN D P, et al. Cryo-EM study of the chromatin fiber reveals a double helix twisted by tetranucleosomal units. Science, 2014, 344: 376-380.
754. CHEN L Y, LIU L. Current progress and prospects of induced pluripotent stem cells. Science in China Series C-Life Sciences, 2009, 52: 622-636.

755. WANG P F, FU J H, MA W Y, et al. A quantification model for apoptosis in mouse embryos in the early stage of fetation. Science in China Series C-Life Sciences, 2009, 52(10): 922-927.

756. LIU H J, HU Y H, WANG H M, et al. A thixotropic molecular hydrogel selectively enhances Flk1 expression in differentiated murine embryonic stem cells. Soft Matter, 2011, 7: 5430-5436.

757. ZHAO Q, FAN J, HONG W, et al. Inhibition of cancer cell proliferation by 5-fluoro-2'-deoxycytidine, a DNA methylation inhibitor, through activation of dna damage response pathway. Springer Plus, 2012, 1: 65.

758. ZHU G Z, LI Y J, ZHU F, et al. Stephen TW, Sun QM, Jin P, Chen DH, Coordination of engineered factors with TET1/2 promotes early-stage epigenetic modification during somatic cell reprogramming. Stem Cell Reports, 2014, 2: 253-261.

759. LIU Q, WANG G, CHEN Y, et al. A miR-590/Acvr2a/ Rad51b axis regulates dna damage repair during mESC Proliferation. Stem Cell Reports, 2014, 3: 1103-1117.

760. YUE W, LI Y Y, ZHANG T, et al. ESC-Derived basal forebrain cholinergic neurons ameliorate the cognitive symptoms associated with alzheimer's disease in mouse models. Stem Cell Reports, 2015, 5(5): 776-790.

761. CHI L, FAN B, ZHANG K, et al. Targeted differentiation of regional ventral neuroprogenitors and related neuronal subtypes from human pluripotent stem cells. Stem Cell Reports, 2016, 7: 941-954.

762. LIU Z, HUI Y, SHI L, et al. Efficient CRISPR/Cas9-mediated versatile, predictable, and donor-free gene knockout in human pluripotent stem cells. Stem Cell Reports, 2016, 7: 496-507.

763. WANG F, XIONG L, HUANG X, et al. miR-210 directly suppresses BNIP3 expression to protect against the hypoxia-induced apoptosis of neural progenitor cells. Stem Cell Res, 2013, 11(1): 657-667.

764. YANG D, WANG G, ZHU S, et al. MiR-495 suppresses mesendoderm differentiation of mouse embryonic stem cells via the direct targeting of Dnmt3a. Stem Cell Res, 2014, 12: 550-561.

765. LIU J F, LUO X L, XU Y L, et al. Single-stranded DNA binding protein Ssbp3 induces differentiation of mouse embryonicstem cells into trophoblast-like cells. Stem Cell Res Ther, 2016, 7(1): 79.

766. CHEN T, YUAN D, WEI B, et al. E-cadherin-mediated cell-cell contact is critical for induced pluripotent stem cell generation. Stem Cells, 2010, 28(8): 1315-1325.

767. YUAN X, WAN H, ZHAO X, et al. Brief report: combined chemical treatment

enables Oct4-induced reprogramming from mouse embryonic fibroblasts. Stem Cells, 2011, 29(3): 549-553.
768. GAO Y, WEI J, HAN J, et al. The novel function of oct4b isoform-265 in genotoxic stress. Stem Cells, 2012, 30: 665-672.
769. YE D, WANG G, LIU Y, et al. miR-138 promotes induced pluripotent stem cell generation through the regulation of the p53 signaling. Stem Cells, 2012, 30(8): 1645-1654.
770. LUO M, LING T, XIE W B, et al. NuRD blocks reprogramming of mouse somatic cells into pluripotent stem cells. Stem Cells, 2013, 31: 1278-1286.
771. CHEN Q, GAO S, HE W, et al. Xist repression shows time-dependent effects on the reprogramming of female somatic cells to induced pluripotent stem cells. Stem Cells, 2014, 32(10): 2642-2656.
772. CHEN J, GAO Y, HUANG H, et al. The combination of Tet1 with Oct4 generates high-quality mouse-induced pluripotent stem cells. Stem Cells, 2015, 33(3): 686-698.
773. SHEN J, JIA W, YU Y, et al. Pwp1 is required for the differentiation potential of mouse embryonic stem cells through regulating Stat3 signaling. Stem Cells, 2015, 33(3): 661-673.
774. LU W, FANG L, OUYANG B, et al. Actl6a protects embryonic stem cells from differentiating into primitive endoderm. Stem Cells, 2015, 33(6): 1782-1793.
775. FANG L, ZHANG J, ZHANG H, et al. H3K4 methyltransferase set1a is a key Oct4 coactivator essential for generation of Oct4 positive inner cell mass. Stem Cells, 2016, 34(3): 565-580.
776. ZOU K, HOU L, SUN K, et al. Improved efficiency of female germline stem cell purification using fragilis-based magnetic bead sorting. Stem Cells Dev, 2011, 20: 2197-2204.
777. LIU Z, WAN H, WANG E, et al. Induced pluripotent stem induced cells show better constitutive heterochromatin remodeling and developmental potential after nuclear transfer than their parental cells. Stem Cells Dev, 2012, 21(16): 3001-3009.
778. WANG P, ZHANG H L, LI W, et al. Generation of patient-specific induced neuronal cells using a direct reprogramming strategy. Stem Cells Dev, 2014, 23: 16-23.
779. WEI T, JIA W, QIAN Z, et al. Folic acid supports pluripotency and reprogramming by regulating LIF/STAT3 and MAPK/ERK signaling. Stem Cells Dev, 2017, 26(1): 49-59.
780. YANG Y, ZHANG X, YI L, et al. Naïve induced pluripotent stem cells generated from β-thalassemia fibroblasts allow efficient gene correction with CRISPR/Cas9.

Stem Cells Transl Med, 2016, 5(2): 267.

781. ZHANG X L, ZHANG Q, XIN Q, et al. Complex structures of the abscisic acid receptor PYL3/RCAR13 reveal a unique regulatory mechanism. Structure, 2012, 20(5): 780-790.

782. HONG H, CAI Y, ZHANG S, et al. Molecular basis of substrate specific acetylation by N-terminal acetyltransferase NatB. Structure, 2017, 25(4): 641-649. e3.

783. DENG C Y, LI J M, MA W Y. Detection of FRET efficiency in imaging systems by the photo-bleaching of acceptor. Talanta, 2010, 82(2): 771-774.

784. FENG X, LIU X, ZHANG W, et al. p53 directly suppresses BNIP3 expression to protect against hypoxia-induced cell death. EMBO J, 2011, 30(16): 33097-33415.

785. ZHOU P, SHA H, ZHU J. The role of T-helper 17 (Th17) cells in patients with medulloblastoma. J Int Med Res, 2010, 38: 611-619.

786. YING X B, DONG L, ZHU H, et al. RNA-dependent RNA polymerase 1 from Nicotiana tabacum suppresses RNA silencing and enhances viral infection in Nicotiana benthamiana. Plant Cell, 2010, 22: 1358-1372.

787. ZHANG Z, ZHANG S, ZHANG Y, et al. Arabidopsis floral initiator SKB1 confers high salt tolerance by regulating transcription and pre-mRNA splicing through altering histone H4R3 and small nuclear ribonucleoprotein LSM4 methylation. Plant Cell, 2011, 23: 396-411.

788. LIU X G, KIM Y J, MUELLER R, et al. Terminates floral stem cell maintenance in arabidopsis by directly repressing WUSCHEL through recruitment of polycomb group proteins. Plant Cell, 2011, 23: 3654-3670.

789. ZHANG Z H, CHEN H, HUANG X H, et al. BSCTV C2 attenuates the degradation of SAMDC1 to suppress DNA methylation-mediated gene silencing in arabidopsis. Plant Cell, 2011, 23: 273-288.

790. GAO J, ZHU Y, ZHOU W, et al. NAP1 family histone chaperones are required for somatic homologous recombination in Arabidopsis. Plant Cell, 2012, 24: 1437-1447.

791. WANG L L, SONG X W, GU L F, et al. NOT2 proteins promote polymerase II–dependent transcription and interact with multiple microRNA biogenesis factors in arabidopsis. Plant Cell, 2013, 25: 715-727.

792. ZHU Y, RONG L, LUO Q, et al. The histone chaperone NRP1 forms protein complex with WEREWOLF in promoting GLABRA2 transcription for proper Arabidopsis root hair development. Plant Cell, 2017, 2: 260-276.

793. LIU L J, ZHANG Y Y, TANG S Y, et al. An efficient system to detect protein ubiquitination by agroinfiltration in Nicotiana benthamiana. Plant J, 2010, 61: 893-

903.

794. LI D, LIU M, FANG Y Y, et al. DRD1-Pol V-dependent self-silencing of an exogenous silencer restricts the non-cell autonomous silencing of an endogenous target gene. Plant J, 2011, 68: 633-645.

795. ZHU Y, WENG M, YANG Y, et al. Arabidopsis homologues of the histone chaperone ASF1 are crucial for chromatin replication and cell proliferation in plant development. Plant J, 2011, 66: 443-455.

796. YANG L P, FANG Y Y, AN C P, et al. C2-mediated decrease in DNA methylation, accumulation of siRNAs, and increase in expression for genes involved in defense pathways in plants infected with beet severe curly top virus. Plant J, 2013, 73: 910-917.

797. FAN H, ZHANG Z, WANG N, et al. SKB1/PRMT5-mediated histone H4R3 dimethylation of Ib subgroup bHLH genes negatively regulates iron homeostasis in Arabidopsis thaliana. Plant J, 2014, 77: 209-221.

798. XU C S, WU X, ZHU J H. VEGF Promotes proliferation of human glioblastoma multiforme stem-like cells through VEGF receptor 2. Scientific World J, 2013, 2013(5): 417413.

799. ZHAO J, SUN H Q, DENG W Q, et al. Piwi-like 2 mediates fibroblast growth factor signaling during gastrulation of zebrafish embryo. Tohoku J Exp Med, 2010, 222(1): 63-68.

800. CHEN X L, WEI S S, MA Y, et al. Quantitative proteomics analysis identifies mitochondria as therapeutic target of multidrug-resistance in ovarian cancer. Theranostics, 2014, 4(12): 1164-1175.

801. TANG M, LU X, ZHANG C, et al. Downregulation of SIRT7 by 5-fluorouracil induces radiosensitivity in human colorectal cancer. Theranostics, 2017, 7(5): 1346-1359.

802. LIU Y, JIANG M, LI C, et al. Human t-complex protein 11 (TCP11), a testis-specific gene product, is a potential determinant of the sperm morphology. Tohoku J Exp Med, 2011, 224(2): 111-117.

803. LI Y, ZHAO D, CHEN Z, et al. YEATS domain: Linking histone crotonylation to gene regulation. Transcription, 2017, 8: 9-14.

804. YUE C M, JING N H. The promise of stem cells in the therapy of Alzheimer's disease. Transl Neurodegener, 2015, 4: 8-12.

805. GOU L T, DAI P, LIU M F. Small noncoding RNAs and male infertility. WIREs RNA, 2014, 5(6): 733-745.

806. SU S, ZHANG M, LI L, et al. Polycomb group genes as the key regulators in gene silencing. Wuhan University Journal of Natural Science, 2014, 19: 1-7.
807. LI D, SUN H, DENG W, et al. Zili antagonizes Bmp signaling to regulate dorsal-ventral patterning during zebrafish early embryogenesis. Zoolog Sci, 2011, 28(6): 397-402.

附录 2　获得的国家科学技术奖项目

附表 1　"细胞编程与重编程的表观遗传机制"获得的国家科学技术奖项目

项目批准号	获奖项目名称	完成人（排名）[1]	完成单位	获奖项目编号	获奖类别[2]	获奖等级	获奖时间
91319306	哺乳动物多能性干细胞的建立与调控机制研究	周琪（1）高绍荣（4）	中国科学院动物研究所	2014-Z-105-2-01-R04	Z	二等奖	2014 年
91219201	乳腺癌发生发展的表观遗传机制	尚永丰（1）石磊（3）	北京大学	2016-Z-106-2-02	Z	二等奖	2016 年
91019009	食管癌规范化治疗关键技术的研究及应用推广	刘芝华（5）	中国医学科学院肿瘤医院	2013-J-25301-1-01-R05	J	一等奖	2013 年
90919044	白血病表观遗传学基础及临床应用研究	于力（1）	中国人民解放军总医院	2009-J-233-2-07-R01	J	二等奖	2009 年
90919024	基因的故事：解读生命密码	陈润生（1）	中国科学院生物物理研究所	2013-J-204-2-02	J	二等奖	2013 年
90919002	脑组织修复重建和干细胞示踪技术及转化应用	朱剑虹（1）	复旦大学附属华山医院	2014-J-25302-2-03-R01	J	二等奖	2014 年
91019002	基因工程小鼠等相关疾病模型研发与应用	杨中州（3）	南京大学	2016-J-235-2-02-R03	J	二等奖	2016 年

注：1. 承担项目的专家获得：国家自然科学二等奖 2 项，均为第一完成人；国家科技进步奖一等奖 1 项，为第五完成人；国家科学技术进步奖二等奖 4 项，其中 3 项是第一完成人，1 项是第三完成人。有 4 项研究成果入选"中国科学十大进展"。表中只列出了与该重大研究计划资助项目有关的完成人，括号中为其排名顺序。

2. "Z"代表国家自然科学奖，"J"代表国家科学技术进步奖。

附录3 代表性发明专利

附表2 "细胞编程与重编程的表观遗传机制"已获授权的代表性发明专利

项目批准号	发明名称	发明人（排名）*	专利号	专利申请时间	专利权人	授权时间
90919024	Method for diagnosing bladder cancer by analyzing DNA methylation profiles in urine sediments and its kit	朱景德（1）	US2010/0317000A1	2009年6月	南京大学	2010年6月
90919059	Use of fenretinide or bioactive derivatives thereof and pharmaceutical compositions comprising the same	张济（1）	US8088822	2008年7月	张济	2012年1月
90919015	利用甲基化特异性荧光法检测p16基因CpG岛甲基化的引物组	邓大君（1）、国静（2）	US9416423B2	2011年1月	北京大学	2016年8月
90919005	人黑色素瘤细胞特异表达的长非编码RNA序列及用途	宋旭（1）	200910059953.4	2009年7月	四川大学	2011年4月
90919005	人黑色素瘤细胞相关的长非编码RNA及其用途	宋旭（1）	200910059954.9	2009年7月	四川大学	2011年4月
90919003/91519308	一种调控植物生长发育的方法及其应用	董爱武（2）	200910047986.7	2009年3月	复旦大学	2012年11月
90919042	一种融合蛋白TAT-NANOG及其编码基因与应用	戴建武（1）	201110084055.1	2011年4月	中国科学院遗传与发育生物学研究所	2013年1月
90919003/91519308	水稻花期干旱胁迫响应蛋白及其编码基因与应用	董爱武（4）	201110402703.3	2011年12月	复旦大学	2013年1月
90917019	一种检测p52Shc/p66Shc基因表达模式的分子诊断试剂盒及其检测方法及应用	刘喆（1）	201210066328.4	2012年3月	天津医科大学	2013年11月
90919005	人黑色素瘤细胞相关的长非编码RNA的RNA干扰靶点RNA及用途	宋旭（1）	201010544055.0	2009年7月	四川大学	2013年4月
90919060	诱导型神经干细胞的制备方法	周琪（1）	201101164400.2	2011年6月	中国科学院动物研究所	2014年12月
90919042	一种融合蛋白TAT-OCT4及其编码基因与应用	戴建武（1）	201110084049.6	2011年4月	中国科学院遗传与发育生物学研究所	2014年3月
90919002	一种从人皮肤细胞直接诱导号神经元细胞及其制备方法	朱剑虹（1）	201110140481.2	2011年5月	复旦大学附属华山医院	2015年2月

续表

项目批准号	发明名称	发明人（排名）*	专利号	专利申请时间	专利权人	授权时间
91019019	一种用于儿童急性淋巴细胞白血病基因分型的试剂盒	韩敬东(2)	201410250033.1	2014年6月	首都医科大学附属北京儿童医院；中国科学院上海生命科学研究院计算生物学研究所；宁波海尔施基因科技有限公司	2015年7月
91219303	从多潜能干细胞诱导胆碱能神经元的方法	景乃禾(1)	201210056400.5	2012年3月	中国科学院上海生命科学研究院	2015年6月
91319310	孤雄单倍体胚胎干细胞系及其制法和应用	李劲松(1)	201310103236.3	2013年3月	中国科学院上海生命科学研究院	2015年9月
91019022	用于诱导多能干细胞的融合蛋白及其使用方法	陈大华(1)	201410020902.1	2014年1月	中国科学院动物研究所	2016年4月
91019020	端粒结合蛋白DAXX在制备肿瘤细胞调控剂中的应用	松阳洲(1)，黄军就(2)	201305273.5	2013年1月	中山大学	2016年4月
91019024	一种连续低剂量辐射富集肿瘤干细胞的方法	孙英丽(1)	201410367053.7	2014年7月	中国科学院北京基因组研究所	2016年8月
90919055	鉴定骨髓中是否存在保护白血病干细胞微环境的试剂盒及其应用	洪登礼(1)	201410141003.7	2014年4月	上海交通大学医学院	2016年9月
91519325	5-醛基胞嘧啶特异性化学标记方法及相关应用	伊成器(1)	201410486471.8	2014年9月	北京大学	2016年9月
90919017	抑制肿瘤转移的方法和试剂	张笑人(1)	201210380158.7	2012年1月	中国科学院上海生命科学研究院	2016年5月
91519331	基因IKZF3在制备肺癌诊断试剂盒中的应用及试剂盒	刘喆(1)	201410056352.9	2014年2月	天津医科大学	2016年8月
91519319	一种含人Rad51C启动子的载体及其应用	毛志勇(3)	201310415502.6	2013年9月	同济大学	2016年8月
91419309	一种调控诱导产生的视网膜前体细胞的方法	金颖(1)	201210279963.0	2012年8月	上海生命科学院	2016年9月

续表

项目批准号	发明名称	发明人（排名）*	专利号	专利申请时间	专利权人	授权时间
91519332	一种提高蜂王浆性能的方法	陈忠周	2015093421.1	2015年12月	中国农业大学	
91519333	一种基于针对特定基因反向互补序列和Py位点调可变剪切的方法	单革（1）	201610852407.6	2016年9月	中国科学技术大学	
91319306	生长因子诱导生成神经前体细胞的方法	高睿（1）	201611097574.0	2016年	同济大学	
91319306	有助于诱导生成神经前体细胞的诱导肽基	高睿（1）	201611101061.2	2016年	同济大学	
91319306	利用细胞表面分子ALPPL2鉴定人类naive多态能性的方法	王译萱（1）	201611049238.9	2016年	同济大学	
91419510	改变细胞命运的手段	裴端卿(1), 陈可实(3)	201610616149.1	2016年	中国科学院广州生物医药与健康研究院	
90919017	一种CD40胞外区的表达纯化及其用途	张笑人（1）	201410680486.8	2014年11月	中国科学院上海生命科学研究院	
90919017	利用microRNA海绵技术预防和治疗肿瘤的方法	张笑人（1）	201210295556.9	2012年8月	中国科学院上海生命科学研究院	2017年5月
91019022	用于诱导性多能干细胞的融合蛋白及其使用方法	陈大华（1）	201410020902.1	2014年1月	中国科学院动物研究所	2017年2月
91019022	用于诱导性多能干细胞的融合蛋白及其使用方法	陈大华（1）	PCT/CN2014/080042	2014年6月	中国科学院动物研究所	
91319308	miRNA对m6A修饰水平的调控方法及其应用	周琪(1), 杨运桂(2), 王秀杰(3)	PCT/CN2015/071582	2015年1月	中国科学院动物研究所，中国科学院北京基因组研究所，中国科学院遗传与发育生物学研究所	

注：*只列举了与本重大研究计划资助项目有关的发明人，括号中为排名顺序。

附录 4　人才队伍培养与建设情况

本重大研究计划的成功实施带动了我国表观遗传研究水平的全面提升和跨越式发展，吸引和培养了一支结构合理、爱岗敬业、创新能力突出、具有国际竞争力的研究队伍，在国际学术界产生了重大影响。

1　研究队伍创新能力突出

参与本重大研究计划的课题负责人中，134 位具有正高级职称，16 位具有副高级职称。研究队伍情况详见附表 3。

2　中青年学术骨干培养成就斐然

本重大研究计划注重中青年学术骨干的培养，为其提供优良的研究与成长环境。多名学术骨干在项目期间取得了重大学术突破，获得了多项国际、国内荣誉，我国表观遗传研究整体实力和国际影响力明显增强（详见附表 4）。项目执行期间在人才培养方面的重要成果包括如下。

（1）项目负责人及指导专家组专家共 4 人当选中国科学院院士：尚永丰（2009）、徐国良（2015）、周琪（2015）和曹晓风（2015）。

（2）15人获得国家自然科学基金委员会"国家杰出青年科学基金"资助：许瑞明（2009）、杨洁（2011）、惠利建（2012）、李劲松（2012）、刘默芳（2013）、高绍荣（2013）、徐彦辉（2014）、刘江（2014）、朱冰（2014）、李国红（2015）、王平（2016）、杨运桂（2016）、单革（2017）、李海涛（2017）和颉伟（2017）。

（3）16人获得国家自然科学基金委员会"优秀青年科学基金"资助：赵颖（2011）、蔡军（2012）、娄继忠（2012）、陈忠周（2012）、汪小我（2013）、王艳（2013）、王强（2013）、颉伟（2014）、孙露洋（2014）、慈维敏（2014）、伊成器（2015）、李伟（2015）、陈捷凯（2015）、徐麟（2015）、毛志勇（2016）和陈凌懿（2017）。

（4）2人入选"千人计划"：孙毅（2010）和丁胜（2015）。

（5）1人入选"青年千人计划"：陆发隆（2016）。

（6）8人入选"万人计划"科技创新领军人才：周琪（2012）、刘江（2016）、陈大华（2016）、惠利建（2016）、王秀杰（2016）、高绍荣（2016）、朱冰（2016）和李劲松（2016）。

（7）2人入选"万人计划"青年拔尖人才：薛宇（2014）和李伟（2014）。

（8）8人入选科学技术部"中青年科技创新领军人才"：康九红（2013）、李劲松（2013）、王艳（2014）、陈大华（2014）、王秀杰（2014）、高绍荣（2015）、常永生（2015）和单革（2016）。

（9）3人入选教育部"长江学者奖励计划"特聘教授：高绍荣（2014）、曾木圣（2014）和徐彦辉（2015）。

（10）1人入选教育部"长江学者奖励计划"青年项目：王艳（2016）。

（11）1人带头的团队入选教育部"创新团队发展计划"：康九红（2012）。

（12）8人入选教育部"新世纪优秀人才支持计划"：薛宇（2009）、蔡军（2011）、吴昊（2011）、王艳（2012）、陈凌懿（2013）、刘喆（2013）、石磊（2013）和常永生（2013）。

（13）5人入选"新世纪百千万人才工程"国家级人选：周金秋（2009）、韩敬东（2011）、王艳（2014）、李劲松（2015）和高绍荣（2015）。

（14）5人获谈家桢生命科学创新奖：徐国良（2012）、周琪（2013）、李劲松（2014）、徐彦辉（2015）和高绍荣（2015）。

（15）2人获何梁何利基金"科学与技术创新奖"：周琪（2010）和李劲松（2014）。

（16）3人获国务院"政府特殊津贴"：康九红（2013）、郭慧珊（2015）和高绍荣（2016）。

（17）4人作为首席科学家带头承担了科技部"973"项目：康九红（2011、2016）、李卫（2016）、高绍荣（2016）和孙方霖（2017）。

多人获其他省部级奖项。

3 后备人才培养卓有成效

本重大研究计划特别重视博士后、博士生等青年后备人才的培养工作。项目结束时，出站博士后 84 名，毕业博士生 441 名，毕业硕士生 255 名，在站博士后 48 名，在读博士生 396 名，在读硕士生 84 名，为我国表观遗传基础研究领域原始创新能力的全面提升和可持续发展提供了重要的人才保证。

4 科学家在国际学术界的影响与日俱增

本重大研究计划的实施在国际上引起了广泛关注，产生了重要国际影响。项目成果在国际顶尖学术期刊 *Cell*、*Nature*、*Science* 上发表重要学术论文 20 篇，其中 7 篇被相关机构作为高被引用论文；多篇高水平学术论文获得 *F1000* 的评论和推荐。参与本重大研究计划项目科学家中，如尚永丰、孟安明、徐国良、周琪、朱冰、高绍荣、陈大华、许瑞明、陈晔光、邓宏魁、景乃禾、李国红、陈晔光、惠利健、翁杰敏、张奇伟、裴端卿、李海涛、郭惠珊等，有50余位科学家分别担任 *Cell*、*Cell Stem Cell*、*Cell Research*、*Biomaterials*、*Journal of Biological Chemistry*、*Aging Cell*、*Journal of Molecular Cell Biology*、*EMBO Reports*、*Genes & Development*、*BMC Genomics* 等国际主流学术刊物的副主编或编委等。在项目执行期间，本重大研究计划的多位项目负责人应邀担任国际学术会议的主席和分会主持人，项目承担人在国际重要学术会议做特邀报告261人次。

附表3 "细胞编程与重编程的表观遗传机制"研究队伍情况

项目负责人（按职称统计）	承担项目数/项				经费/万元	经费占总经费百分比/%
	培育项目	重点支持项目	集成项目	小计		
正高	55	23	56	134	16228	88.8
副高	13	0	3	16	1340	7.3
中级	0	0	0	0	0	0
博士后	0	0	0	0	0	0
其他	0	0	0	0	0	0
合计	68	23	59	150	17568	96.2

附表4 "细胞编程与重编程的表观遗传机制"中青年学术带头人培养情况

项目批准号	姓名	单位	所获人才计划项目名称或荣誉	时间
91219201	尚永丰	天津大学	中国科学院院士	2009年
91319308	周琪	中国科学院动物研究所	中国科学院院士	2015年
90919061	徐国良	中国科学院上海生命科学研究院	中国科学院院士	2015年
90919000	曹晓风	中国科学院遗传与发育生物学研究所	中国科学院院士	2015年
90919029	许瑞明	中国科学院生物物理研究所	国家杰出青年科学基金	2009年
90919032	杨洁	天津医科大学	国家杰出青年科学基金	2011年
91319307	惠利健	中国科学院上海生命科学研究院	国家杰出青年科学基金	2012年
91319310	李劲松	中国科学院上海生命科学研究院	国家杰出青年科学基金	2012年
9141930035	刘默芳	中国科学院上海生命科学研究院	国家杰出青年科学基金	2013年
91319306	高绍荣	同济大学	国家杰出青年科学基金	2013年
91419305	朱冰	中国科学院生物物理研究所	国家杰出青年科学基金	2014年
91419301	徐彦辉	复旦大学	国家杰出青年科学基金	2014年
91219104	刘江	中国科学院北京基因组研究所	国家杰出青年科学基金	2014年
91219202	李国红	中国科学院生物物理研究所	国家杰出青年科学基金	2015年
9151930075	王平	同济大学	国家杰出青年科学基金	2016年
91319308	杨运桂	中国科学院北京基因组研究所	国家杰出青年科学基金	2016年
91519333	单革	中国科学技术大学	国家杰出青年科学基金	2017年

细胞编程与重编程的表观遗传机制

续表

项目批准号	姓名	单位	所获人才计划项目名称或荣誉	时间
91519304	李海涛	清华大学	国家杰出青年科学基金	2017年
91519326	颉伟	清华大学	国家杰出青年科学基金	2017年
91319302	赵颖	北京大学	优秀青年科学基金	2011年
90919054	蔡军	首都医科大学	优秀青年科学基金	2012年
91219103	娄继忠	中国科学院生物物理研究所	优秀青年科学基金	2012年
90919043	陈忠周	中国农业大学	优秀青年科学基金	2012年
91019016	汪小我	清华大学	优秀青年科学基金	2013年
91219201	王艳	天津医科大学	优秀青年科学基金	2013年
90919011	王强	中国科学院动物研究所	优秀青年科学基金	2013年
91519326	颉伟	清华大学	优秀青年科学基金	2014年
91219201	孙露洋	天津大学	优秀青年科学基金	2014年
91519307	慈维敏	中国科学院北京基因组研究所	优秀青年科学基金	2014年
91519325	伊成器	北京大学	优秀青年科学基金	2015年
91319308	李伟	中国科学院动物研究所	优秀青年科学基金	2015年
91419310	陈捷凯	中国科学院广州生物医药与健康研究院	优秀青年科学基金	2015年
91419302	徐麟	中国科学院上海生命科学研究院植物生理生态研究所	优秀青年科学基金	2015年
91519319	毛志勇	同济大学	优秀青年科学基金	2016年
90919009	陈凌懿	南开大学	优秀青年科学基金	2017年
91319309	孙毅	同济大学/附属同济医院	千人计划	2010年
91519318	丁胜	清华大学	千人计划	2015年
90919033	陆发隆	中科院遗传与发育生物学研究所	青年千人计划	2016年
90919060	周琪	中国科学院动物研究所	"万人计划"科技创新领军人才	2012年
91519306	刘江	中国科学院北京基因组研究所	"万人计划"科技创新领军人才	2016年
91019022	陈大华	中国科学院动物研究所	"万人计划"科技创新领军人才	2016年
91319307	惠利健	中国科学院上海生命科学研究院	"万人计划"科技创新领军人才	2016年
91319308	王秀杰	中国科学院遗传与发育生物学研究所	"万人计划"科技创新领军人才	2016年
91319306	高绍荣	同济大学	"万人计划"科技创新领军人才	2016年
91419305	朱冰	中国科学院生物物理研究所	"万人计划"科技创新领军人才	2016年
91319310	李劲松	中国科学院上海生命科学研究院	"万人计划"科技创新领军人才	2016年

续表

项目批准号	姓名	单位	所获人才计划项目名称或荣誉	时间
90919001	薛宇	华中科技大学	"万人计划"青年拔尖人才	2014年
91319308	李伟	中国科学院动物研究所	"万人计划"青年拔尖人才	2014年
91219305	康九红	同济大学	中青年科技创新领军人才计划	2013年
91319310	李劲松	中国科学院上海生命科学研究院	中青年科技创新领军人才计划	2013年
91219201	王艳	天津医科大学	中青年科技创新领军人才计划	2014年
91019022	陈大华	中国科学院动物研究所	中青年科技创新领军人才计划	2014年
91319308	王秀杰	中国科学院遗传与发育生物学研究所	中青年科技创新领军人才计划	2014年
91319306	高绍荣	同济大学	中青年科技创新领军人才计划	2015年
90919019	常永生	中国医学科学院基础医学研究所	中青年科技创新领军人才计划	2015年
91519333	单革	中国科学技术大学	中青年科技创新领军人才计划	2016年
91319306	高绍荣	同济大学	"长江学者奖励计划"特聘教授	2014年
91019015	曾木圣	中山大学	"长江学者奖励计划"特聘教授	2014年
91419301	徐彦辉	复旦大学	"长江学者奖励计划"特聘教授	2015年
91219201	王艳	天津医科大学	"长江学者奖励计划"青年项目	2016年
90919028	康九红	同济大学	"创新团队发展计划"带头人	2012年
90919001	薛宇	华中科技大学	新世纪优秀人才支持计划	2010年
90919054	蔡军	首都医科大学	新世纪优秀人才支持计划	2011年
91019013	吴旻	武汉大学	新世纪优秀人才支持计划	2011年
91219201	王艳	天津医科大学	新世纪优秀人才支持计划	2012年
90919009	陈凌懿	南开大学	新世纪优秀人才支持计划	2013年
91019012	刘喆	天津医科大学	新世纪优秀人才支持计划	2013年
91219102	石磊	天津医科大学	新世纪优秀人才支持计划	2013年
90919019	常永生	中国医学科学院基础医学研究所	新世纪优秀人才支持计划	2013年
90919027	周金秋	中国科学院上海生命科学研究院	"新世纪百千万人才工程"国家级人选	2009年
91019019	韩敬东	中国科学院上海生命科学研究院	"新世纪百千万人才工程"国家级人选	2011年
91219201	王艳	天津医科大学	"新世纪百千万人才工程"国家级人选	2014年
91319310	李劲松	中国科学院上海生命科学研究院	"新世纪百千万人才工程"国家级人选	2015年
91319306	高绍荣	同济大学	"新世纪百千万人才工程"国家级人选	2015年

细胞编程与重编程的表观遗传机制

续表

项目批准号	姓名	单位	所获人才计划项目名称或荣誉	时间
90919061	徐国良	中国科学院上海生命科学研究院	谈家桢生命科学创新奖	2012 年
91319308	周琪	中国科学院动物研究所	谈家桢生命科学创新奖	2013 年
91319310	李劲松	中国科学院上海生命科学研究院	谈家桢生命科学创新奖	2014 年
91419301	徐彦辉	复旦大学	谈家桢生命科学创新奖	2015 年
91319306	高绍荣	同济大学	谈家桢生命科学创新奖	2015 年
91319308	周琪	中国科学院动物研究所	何梁何利基金"科学与技术创新奖"	2010 年
91319310	李劲松	中国科学院上海生命科学研究院	何梁何利基金"科学与技术创新奖"	2014 年
91219305	康九红	同济大学	政府特殊津贴	2013 年
91519327	郭慧珊	中国科学院微生物研究所	政府特殊津贴	2015 年
91319306	高绍荣	同济大学	政府特殊津贴	2016 年
90919028	康九红	同济大学	"973"项目首席科学家	2011 年
91519320	康九红	同济大学	"973"项目首席科学家	2016 年
91319306	高绍荣	同济大学	"973"项目首席科学家	2016 年
91419304	孙方霖	同济大学	"973"项目首席科学家	2017 年
91519317	李卫	中国科学院动物研究所	"973"项目首席科学家	2016 年
91019012	刘喆	天津医科大学	中国侨界创新人才奖	2016 年
91519324	唐铁山	中国科学院动物研究所	中国侨界创新人才奖	2016 年
91419301	徐彦辉	复旦大学	药明康德生命化学学者奖	2015 年
91519325	伊成器	北京大学	药明康德生命化学研究奖	2016 年
91319306	高绍荣	同济大学	明治生命科学奖	2015 年
91519315	程涛	中国医学科学院	全国优秀科技工作者	2016 年
91419301	徐彦辉	复旦大学	中国优秀青年科技人才	2016 年
91219104	刘江	中国科学院北京基因组研究所	TWAS, Young Affiliate	2015 年
90919061	徐国良	中国科学院上海生命科学研究院	TWAS 2013 年度生物学奖	2013 年
91019009	刘芝华	中国医学科学院肿瘤医院	"国家特支计划"百千万工程领军人才	2013 年
91319310	李劲松	中国科学院上海生命科学研究院	光华工程科技奖	2016 年
90919056	相辉	中国科学院昆明动物研究所	"西部之光"人才项目重点项目	2011 年
91019019	韩敬东	中国科学院上海生命科学研究院	德国马普学会"Max Planck Fellow"	2011 年

续表

项目批准号	姓名	单位	所获人才计划项目名称或荣誉	时间
91519317	赵小阳	中国科学院动物研究所	国际转基因协会（ISTT）首届青年科学家奖	2011年
91019020	黄军就	中山大学	Nature 2015年全球十大科学人物	2015年
91019020	黄军就	中山大学	全国干细胞协会"干细胞青年研究员奖"	2016年
91419301	徐彦辉	复旦大学	树兰医学青年奖	2015年
90919020	吴际	上海交通大学	上海市优秀学术带头人	2010年
90919028	康九红	同济大学	上海市优秀学术带头人	2011年
91019019	韩敬东	中国科学院上海生命科学研究院	上海市优秀学术带头人	2011年
91319309	孙毅	同济大学/附属同济医院	上海市优秀学术带头人	2013年
91219302	翁杰敏	华东师范大学	上海市优秀学术带头人	2014年
91219302	吴奇涵	华东师范大学	上海市优秀学术带头人	2014年
91319307	惠利健	中国科学院上海生命科学研究院	上海市优秀学术带头人	2014年
91319306	高绍荣	同济大学	上海市优秀学术带头人	2015年
91519323	章小清	同济大学	上海市优秀学术带头人	2015年
91419306	蓝斐	复旦大学	上海市优秀学术带头人	2016年
90919046	景乃禾	中科院上海生化细胞所	上海市领军人才	2010年
90919061	徐国良	中国科学院上海生命科学研究院	上海市领军人才	2012年
91019019	韩敬东	中国科学院上海生命科学研究院	上海市领军人才	2015年
90919061	徐国良	中国科学院上海生命科学研究院	上海市五一劳动奖章	2013年
91319310	李劲松	中国科学院上海生命科学研究院	上海市五一劳动奖章	2016年
91519309	江赐忠	同济大学	上海市东方学者计划	2012年
91319307	惠利健	中国科学院上海生命科学研究院	上海市青年科技英才	2014年
90919025	翁杰敏	华东师范大学	上海市浦江人才计划	2009年
91019001	文波	复旦大学	上海市浦江人才计划	2011年
91019001	文波	复旦大学	上海市曙光计划	2011年
91519309	江赐忠	同济大学	上海市曙光计划	2011年
91319306	王译萱	同济大学	上海市晨光人才计划	2014年
91319306	高亚威	同济大学	上海市晨光人才计划	2015年
91319306	高亚威	同济大学	上海市启明星人才计划	2016年
91019001	文波	复旦大学	上海优秀青年人才计划	2012年

细胞编程与重编程的表观遗传机制

续表

项目批准号	姓名	单位	所获人才计划项目名称或荣誉	时间
91319306	王译萱	同济大学	上海市扬帆人才计划	2014 年
91319309	孙毅	同济大学/附属同济医院	上海市医学科技奖一等奖	2013 年
90919028	康九红	同济大学	上海市教育系统"科研新星"	2011 年
91519320	康九红	同济大学	上海市育才奖	2016 年
91319310	李劲松	中国科学院上海生命科学研究院	上海市科技系统先进工作者	2014 年
91019012	刘喆	天津医科大学	天津市"131"创新团队带头人	2014 年
90919009	陈凌懿	南开大学	天津市"131"创新型人才培养工程第二层次	2015 年
81322032	王艳	天津医科大学	天津市"131"创新型人才培养工程第一层次	2012 年
91219102	石磊	天津医科大学	天津市"131"创新型人才培养工程第一层次	2015 年
91219102	石磊	天津医科大学	天津市"青年拔尖人才支持计划"	2014 年
90919032	杨洁	天津医科大学	天津市"学科领军人才培养计划"	2012 年
91219201	王艳	天津医科大学	天津市"用三年时间引进千名以上高层次人才计划"	2012 年
90919032	杨洁	天津医科大学	天津市青年科技奖	2012 年
91019012	刘喆	天津医科大学	天津市青年科技奖	2015 年
91219201	王艳	天津医科大学	天津市特聘教授	2016 年
90919032	杨洁	天津医科大学	天津市优秀科技工作者	2012 年
91519315	程涛	中国医学科学院	天津首届杰出人才	2016 年
91019020	黄军就	中山大学	广东省百千万青年特别资助	2014 年
91019020	松阳洲	中山大学	广东省杰出人才（南粤百杰）	2016 年
91019020	松阳洲	中山大学	广东省领军人才	2010 年
91419310	刘晶	中国科学院广州生物医药与健康研究院	广东省省拔尖人才	2016 年
90919039	杨松光	中国科学院华南植物园	广东省优秀青年科研人员	2016 年
90919035	张新华	河北医科大学	河北省青年拔尖人才	2016 年
90919035	郑斌	河北医科大学	河北省有突出贡献的中青年专家	2012 年
91019011	陆军	东北师范大学	吉林省第十批有突出贡献的中青年专家	2012 年
90919006	马用信	四川大学	四川省学术与技术带头人	2013 年
90919013	邹琳	重庆医科大学	重庆市杰出青年基金获得者	2010 年
90919048	张业	中国医学科学院基础医学研究所	协和学者特聘教授	2011 年

续表

项目批准号	姓名	单位	所获人才计划项目名称或荣誉	时间
91419301	徐彦辉	复旦大学	中源协和创新突破奖	2016年
91519325	伊成器	北京大学	中国化学会青年化学奖	2016年
91519317	李卫	中国科学院动物研究所	中国科学院百人计划	2009年
90919039	毛炳宇	中国科学院昆明动物研究所	中国科学院百人计划	2009年
91019024	孙英丽	中国科学院北京基因组研究所	中国科学院百人计划	2011年
91519311	王纲	中科院上海生化细胞所	中国科学院分子细胞科学卓越创新中心核心骨干	2015年
91519317	赵小阳	中国科学院动物研究所	中国科学院杰出科技成就奖	2013年
90919033	刘春艳	中科院遗传与发育生物学研究所	中国科学院青年创新促进会	2012年
90919056	相辉	中国科学院昆明动物研究所	中国科学院青年创新促进会	2011年
91019019	韩敬东	中国科学院上海生命科学研究院	中国科学院特聘研究员	2015年
91519324	唐铁山	中国科学院动物研究所	中国科学院优秀导师奖	2016年
90919061	徐国良	中国科学院上海生命科学研究院	中国科学院优秀导师奖	2012年
91519324	唐铁山	中国科学院动物研究所	中国科学院朱李月华优秀教师	2016年
91219307	刘超培	中国科学院生物物理研究所	中国科学院博士后出站晋升为副研究员	2013年
91519328	金文星	中国科学院生物物理研究所	中国科学院博士后出站晋升为副研究员	2016年
90919029	王明珠	中国科学院生物物理研究所	中国科学院助理研究员晋升为副研究员	2011年
91519333	陈亮	中国科学技术大学	中科院"百篇优博"	2016年
91519324	王翘楚	中国科学院动物研究所	中国科学院院长特别奖/中国科学院优秀学生标兵	2016年
90919061	郭帆	中国科学院上海生命科学研究院	中科院朱李月华优秀博士奖	2012年
90919061	郭帆	中国科学院上海生命科学研究院	中科院BHPB奖学金	2012年

索 引
（按拼音排序）

DNA 甲基化	12
表观遗传	1
表观遗传学图谱	45
单倍体干细胞	28
非编码 RNA	41
核小体	25
克隆	30
染色质高级结构	42
细胞重编程	2
细胞分化	4
诱导性多能干细胞	28
转分化	17
组蛋白修饰	45
组蛋白变体	26

图书在版编目（CIP）数据

细胞编程与重编程的表观遗传机制 / 细胞编程与重编程的表观遗传机制项目组编． — 杭州：浙江大学出版社，2018.12

ISBN 978-7-308-18869-2

Ⅰ.①细… Ⅱ.①细… Ⅲ.①细胞工程－表现遗传学 Ⅳ.①Q813

中国版本图书馆 CIP 数据核字（2018）第 293805 号

细胞编程与重编程的表观遗传机制

细胞编程与重编程的表观遗传机制项目组　编

丛书统筹	国家自然科学基金委员会科学传播中心
策划编辑	徐有智　许佳颖
责任编辑	金　蕾
责任校对	汪志强　陈静毅
封面设计	程　晨
出版发行	浙江大学出版社
	（杭州市天目山路 148 号　邮政编码 310007）
	（网址：http://www.zjupress.com）
排　　版	杭州中大图文设计有限公司
印　　刷	浙江印刷集团有限公司
开　　本	710mm×1000mm　1/16
印　　张	9.75
字　　数	126 千
版 印 次	2018 年 12 月第 1 版　2018 年 12 月第 1 次印刷
书　　号	ISBN 978-7-308-18869-2
定　　价	88.00 元

版权所有　翻印必究　印装差错　负责调换

浙江大学出版社市场运营中心联系方式（0571）88925591;http://zjdxcbs.tmall.com